Contents

D0258705

Heating houses

Energy flow

- **Energy**, in the form of heat, flows from a warmer to a colder body. When energy flows away from a warm object, the **temperature** of that object decreases.

Measuring temperature

- A **thermogram** uses colour to show temperature; hottest areas are white/yellow, coldest are black/dark blue/purple.

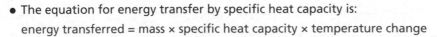

hot end   cold end

Typical range of colours used in a thermogram

- Temperature is a measurement of hotness on an arbitrary scale. You do not need to use a thermometer. It allows one object to be compared to another.

- When the temperature of a body increases, the average **kinetic energy** of the particles increases.

- Heat is a measurement of internal energy. It is measured on an absolute scale.

Specific heat capacity

- All substances have a property called **specific heat capacity**, which is:
  - the energy needed to raise the temperature of 1 kg by 1 °C
  - measured in **joules** per kilogram degree Celsius (J/kg °C) and differs for different materials.

- When an object is heated and its temperature rises, energy is transferred.

- The equation for energy transfer by specific heat capacity is:

  energy transferred = mass × specific heat capacity × temperature change

> Calculate the energy transferred when 30 kg of water cools from 25 °C to 5 °C.
> energy transferred = 30 × 4200 × (25 − 5) = 30 × 4200 × 20
> = 2 520 000 J or 2520 kJ

Specific latent heat

- **Specific latent heat** is:
  - the energy needed to melt or boil 1 kg of the material
  - measured in joule per kilogram (J/kg) and differs for different materials and each of the changes of state.

- When an object is heated and it changes state, energy is transferred, but the temperature remains constant.

- The equation for energy transfer by specific latent heat is:

  energy transferred = mass × specific latent heat

> Calculate the energy transferred when 2.5 kg of water changes from solid to liquid at 0 °C
> energy transferred = 2.5 × 340 000
> = 850 000 J or 850 kJ

In a thermogram, white, yellow and red represent the hottest areas. Black, dark blue and purple represent the coldest areas.

- When a substance changes state, energy is needed to break the bonds that hold the molecules together. This explains why there is no change in temperature.

Improve your grade

Specific heat capacity

Ed uses a stainless steel saucepan to heat his soup from 17 °C to 94 °C. The saucepan has a mass of 1.1 kg and a specific heat capacity of 510 J/kg °C. Energy is required to heat the soup.

(a) Calculate the extra energy required to raise the temperature of the saucepan.
   AO2 [2 marks]

(b) Ed reads that a 1.1 kg copper saucepan will be more energy efficient. Explain why.
   AO2 [2 marks]

P1 Energy for the home

Collins Revision

Revision

New GCSE
Physics

Revision Guide

Higher

For OCR Gateway B

Authors: Chris Sherry
Maureen Elliot

About this book

This book covers the content you will need to revise for GCSE Physics for OCR Gateway B at Higher Level. Written by GCSE examiners, it is designed to help you get the best grade in your GCSE Physics exams.

The book is divided into six modules.

Revision and practice

The book is divided into two parts: **Revision guide** and **Workbook**. Begin by revising a topic in the Revision guide section, then test yourself by answering the exam-style questions for that topic in the Workbook section.

Revision guide

The pages in the Revision guide summarise the content of the exam specification and act as a memory jogger. There is a question (**Improve your grade**) on each page that will help you to check your progress. Typical answers to these questions, and examiner's comments, are provided at the end of the Revision guide section (pages 58–63) for you to compare with your responses. This will help you to improve your answers in the future.

At the end of each module, you will find a **Summary** page. This highlights some important facts from each module.

Workbook

The **Workbook** (pages 74–127) allows you to work at your own pace on some typical exam-style questions. You will find that the actual GCSE questions are more likely to test knowledge and understanding across topics. However, the aim of the Revision guide and Workbook is to guide you through each topic so that you can identify your areas of strength and weakness.

The Workbook also contains example questions that require longer answers (**Extended response questions**). You will find one question that is similar to these in each section of your written exam papers. The quality of your written communication will be assessed when you answer these questions in the exam, so practise writing longer answers, using sentences. For ease of use, the **Answers** to all **Workbook** questions are detachable. These can be found on pages 137–148.

At the end of the Workbook there is a series of **Grade booster checklists** that you can use to tick off the topics when you are confident about them and understand certain key ideas. These Grade boosters give you an idea of the grade at which you are currently working.

Additional features

- **Exam tips** give additional exam advice.
- **Remember boxes** pick out key facts to help you revise.
- A **Glossary** allows quick reference to the definitions of the scientific terms highlighted in bold.

Published by Collins
An imprint of HarperCollins*Publishers*
77–85 Fulham Palace Road
Hammersmith
London W6 8JB

Browse the complete Collins catalogue at:
www.collins.co.uk

© HarperCollinsPublishers 2011

10 9 8 7 6 5 4 3 2

ISBN 978-0-00-741613-4

The authors assert their moral rights to be identified as the authors of this work.

All rights reserved. No part of this publication may be reproduced, stored in a retrieval system, or transmitted in any form or by any means, electronic, mechanical, photocopying, recording or otherwise without the prior written permission of the Publisher or a licence permitting restricted copying in the United Kingdom issued by the Copyright Licencing Agency Ltd., 90 Tottenham Court Road, London W1T 4LP.

British library Cataloguing in Publication Data

A Catalogue record for this publication is available from the British Library.

Written by Chris Sherry and Maureen Elliot.

Project managed by Sally Moon Publishing Services
Design by wired2create
Page make-up by Jordan Publishing Design Limited
Illustrations by Kathy Baxendale, IFA Design Ltd ,
Ken Vail Graphic Design and Nigel Jordan
Edited by Jane Price
Printed and bound by Printing Express, Hong Kong

Acknowledgements

Whilst every effort has been made to trace the copyright holders, in cases where this has been unsuccessful, or if any have inadvertently been overlooked, the Publishers will be pleased to make the necessary arrangements at the first opportunity.

Keeping homes warm

Practical insulation

- Double glazing reduces **energy** loss by conduction. The gap between the two pieces of glass is filled with a gas or contains a **vacuum.**
  - Particles in a gas are far apart. It is very difficult to transfer energy. There are no particles in a vacuum so it is impossible to transfer energy by conduction.

- Loft **insulation** reduces energy loss by conduction and convection:
  - warm air in the home rises
  - energy is transferred through the ceiling by conduction
  - air in the loft is warmed by the top of the ceiling and is trapped in the loft insulation
  - both sides of the ceiling are at the same **temperature** so no energy is transferred
  - without loft insulation, the warm air in the loft can move by convection and heat the roof tiles
  - energy is transferred to the outside by conduction.

- Cavity wall insulation reduces energy loss by conduction and convection:
  - the air in the foam is a good insulator
  - the air cannot move by convection because it is trapped in the foam.

- Insulation blocks used to build new homes have shiny foil on both sides to reduce energy transfer by radiation:
  - energy from the Sun is reflected back to keep the home cool in summer
  - energy from the home is reflected back to keep the home warm in winter.

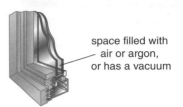

space filled with air or argon, or has a vacuum

A double glazed window

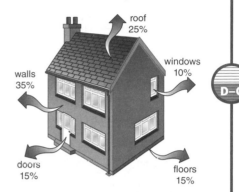

roof 25%

windows 10%

walls 35%

doors 15%

floors 15%

Energy loss from a home

D–C

Remember!
Hot air will only rise into the loft if the loft-hatch is open.

Conduction, convection and radiation

- Energy can be transferred by:
  - conduction – due to the transfer of **kinetic energy** between particles
  - convection – a gas expands when it is heated. This makes it less dense so it rises. The unit of **density** is kg/m^3 or g/cm^3.

$$density = \frac{mass}{volume}$$

  - **radiation** does not need a material to transfer energy. Energy can be transferred through a vacuum.

B–A*

Energy efficiency

$$efficiency = \frac{useful\ energy\ output\ (\times\ 100\ \%)}{total\ energy\ input}$$

- Energy transformations can be shown by Sankey diagrams.
- Energy from the source (home) is lost to the sink (environment).
- Different types of insulation cost different amounts and save different amounts of energy.

$$\textbf{payback time} = \frac{cost\ of\ insulation}{annual\ saving}$$

- Everything that transfers energy will waste some of the energy as heat to the surroundings.
- Buildings that are energy efficient are well insulated; little energy is lost to the surroundings.
- Designers and architects have to make sure that as little energy as possible is wasted.

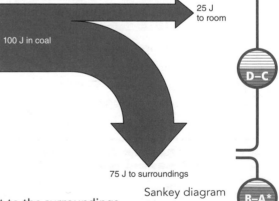

25 J to room

100 J in coal

75 J to surroundings

Sankey diagram

D–C

B–A*

Improve your grade

Energy loss in a cavity wall

The Johnson's house has cavity walls. They decide to have foam injected into the cavity to reduce energy loss.

Explain how energy is transferred to the roof space from the cavity. *AO1* [3 marks]

A spectrum of waves

Wave properties

D–C

- The **amplitude** of a **wave** is the *maximum* displacement of a particle from its rest position.
- The crest of a wave is the *highest point on* a wave above its rest position.
- The trough of a wave is the *lowest point on* a wave *below* its rest position.
- The **wavelength** of a wave is the distance *between* two *successive* points on a wave having the same displacement and moving in the same direction.
- The **frequency** of a wave is the number of complete waves passing a point in one second.
- The equation for the **speed** of a wave is:

  wave speed = frequency × wavelength

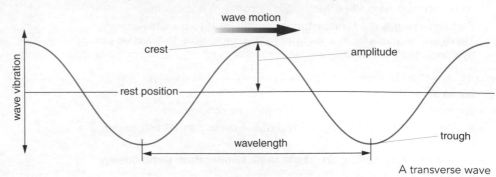

A transverse wave

When Katie throws a stone into a pond, the distance between ripples is 0.3 m and four waves reach the edge of the pond each second.
wave speed = 0.3 × 4 = 1.2 m/s

Remember!
Always give the units in your answer:
– wavelength – metre (m)
– frequency – **hertz** (Hz)
– speed – metre per second (m/s).

B–A*

Microwaves travel at a speed of 300×10^6 m/s. A microwave oven uses microwaves with a frequency of 2.5×10^9 Hz. Calculate the wavelength of the microwaves.

$$\text{Wavelength} = \frac{\text{wave speed}}{\text{frequency}}$$

$$= \frac{300 \times 10^6}{2.5 \times 10^9} = 0.12 \text{ m}$$

Remember!
At higher tier you may be expected to use scientific notation and rearrange equations.

Electromagnetic spectrum

D–C

radio   microwave   infrared   visible   ultraviolet   X-ray   gamma ray

← increasing wavelength   increasing frequency →

Getting messages across

D–C

- Some optical instruments, such as the periscope, use two or more plane mirrors.
- **Refraction** occurs because the speed of waves decreases as the wave enters a more dense medium and increases as the wave enters a less dense medium. The frequency stays the same but the wavelength changes.
- **Diffraction** is the spreading out of a wave as it passes through a gap.
- The size of a communications **receiver** depends on the wavelength of the **radiation**.

Diffraction effects

B–A*

- The amount of diffraction depends on the size of the gap; the most diffraction occurs when the gap is a similar size to the wavelength. Larger gaps show less diffraction.
- Diffraction effects are noticeable in telescopes and microscopes.

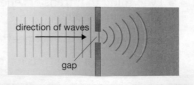

Diffraction at a narrow gap

Improve your grade

Diffraction effects

Light is diffracted as it passes through a narrow slit. Describe how the amount of diffraction depends on the wavelength of the light and the width of the slit. *AO1* [2 marks]

Light and lasers

Morse code

D–C

- The **Morse code** uses a series of dots and dashes to represent letters of the alphabet.
  - This code is used by signalling lamps as a series of short and long flashes of light.
  - It is an example of a **digital signal**.

Sending signals

B–A*

- When a signal is sent by light, electricity, **microwaves** or radio, it is almost instantaneous.
- Each method of transmission has advantages and disadvantages:
  - can the signal be seen by others?
  - can wires be cut?
  - how far does the signal have to travel?

Laser light

B–A*

- White light is made up of different colours of different frequencies out of **phase**.
- **Laser** light has only a single frequency, is in phase and shows low divergence.
- Laser light is used to read from the surface of a compact disc (CD):
  - the surface of the CD is pitted
  - the pits represent the digital signal
  - laser light is shone onto the CD surface and the difference in the reflection provides the information for the digital signal.

Critical angle

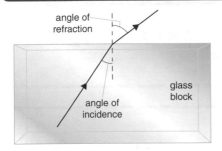
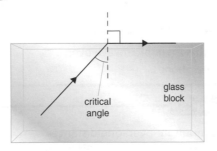
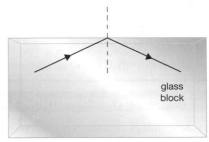
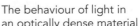

The behaviour of light in an optically dense material

D–C

- When light travels from one material to another, it is normally refracted.
- If it is passing from a more dense material into a less dense, the angle of refraction is larger than the angle of incidence.
- When the angle of refraction is 90°, the angle of incidence is called the **critical angle**.
- If the angle of incidence is bigger than the critical angle, the light is reflected:
  - this is **total internal reflection**.
- Telephone conversations and computer data are transmitted long distances along **optical fibres** at the speed of light (200 000 km/s in glass).
- Some fibres are coated to improve reflection.

Endoscopy

B–A*

- An **endoscope** allows doctors to see inside a body without the need for surgery.
  - Light passes along one set of optical fibres to illuminate the inside of the body.
  - The light is reflected.
  - The reflected light passes up another set of fibres to an eyepiece or camera.

Improve your grade

Sending signals

Adam is standing on top of a hill in line of sight and 10 km away from Becky who is on top of another hill. They can communicate either by using light, radio or electrical signals. Suggest one advantage and one disadvantage of using each type of signal. *AO1* [3 marks]

Cooking with waves

- **Infrared** radiation does not penetrate food very easily.
- **Microwaves** penetrate up to 1 cm into food.
- Microwaves can penetrate glass or plastic but are reflected by shiny metal surfaces:
  – special glass in a microwave oven door reflects microwaves
  – they can cause body tissue to burn.

Electromagnetic spectrum

- Energy is transferred by **waves**:
  – the amount of energy depends on the frequency or wavelength of the wave
  – high frequency (short wavelength) waves transfer more energy.
- Normal ovens cook food by infrared radiation:
  – energy is absorbed by the surface of the food
  – the **kinetic energy** of the surface food particles increases
  – the rest of the food is heated by conduction.
- Microwave ovens cook food by microwave radiation:
  – the water or fat molecules in the outer layers of food vibrate more.

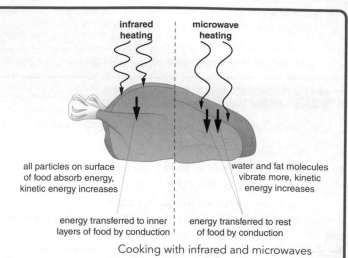

all particles on surface of food absorb energy, kinetic energy increases

water and fat molecules vibrate more, kinetic energy increases

energy transferred to inner layers of food by conduction

energy transferred to rest of food by conduction

Cooking with infrared and microwaves

Microwave properties

- Microwaves have wavelengths between 1 mm and 30 cm.
- Mobile phones use longer wavelengths than microwave ovens.
  – Less energy is transferred by mobile phones.

Microwave communication

- Microwave radiation is used to communicate over long distances.
- The **transmitter** and **receiver** must be in *line of sight*.
  – Aerials are normally situated on the top of high buildings.
- **Satellites** are used for microwave communication.
  – The signal from Earth is received, amplified and re-transmitted back to Earth.
  – Satellites are in line of sight because there are no obstructions in space.
  – Large aerials can handle thousands of phone calls and television channels at once.
- There are concerns about the use of mobile phones and where phone masts are situated.
- Scientists publish results of their studies to allow others to check their findings.

- Signal strength for mobile phones can change a lot over a short distance.
  – Microwaves do not show much **diffraction**.
  – Adverse weather and large areas of water can scatter the signals.
  – The curvature of the Earth limits the line of sight so transmitters have to be on tall buildings or close together.
- Mobile phones can **interfere** with sensitive equipment:
  – They are banned on planes and in many hospitals.

Improve your grade

Microwave transmitters

The Telecom Tower in London is one of the tallest buildings in the city. There are many microwave aerials surrounding the top of the tower.

Explain why they are sited so high up. *AO1* [2 marks]

Data transmission

Digital signals

- **Infrared** signals carry information that allows electronic and electrical devices to be controlled.

- Pressing a button on the remote control device completes the circuit. A coded signal is sent to a **light-emitting diode** or LED at the front of the remote.

- The signal includes a start command, the instruction command, a device code and a stop command.

- The LED transmits the series of pulses. This is received by the device and decoded to allow the television to change channel or volume.

- The switchover from **analogue** to **digital** started in 2009 and is planned to finish by 2015. This may be delayed until more people buy digital radios. The switchover for both radio and TV means:
  - improved signal quality for both picture and sound
  - a greater choice of programmes
  - being able to interact with the programme
  - information services such as programme guides and subtitles.

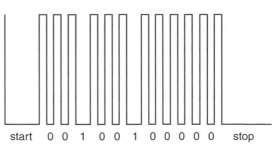

start 0 0 1 0 0 1 0 0 0 0 0   stop

A typical digital signal from a remote

Optical fibres

- **Optical fibres** allow data to be transmitted very quickly using pulses of light.

> **Remember!**
> An optical fibre is solid, not a hollow tube.

Advantages of digital signals

- Before an analogue signal is transmitted, it is added to a carrier **wave**.

- The **frequency** of the carrier wave is usually higher.

- The combined wave is transmitted.

- **Interference** from another wave can also be added and transmitted.

- If the wave is amplified, the interference is amplified as well.

- Interference also occurs on digital signals, but is not apparent because the digital signal only has two values.

- **Multiplexing** allows a large number of **digital signals** to be transmitted at the same time.

Interference on a digital wave

Multiplexing of digital signals

Improve your grade

Advantages of using digital signals and optical fibres

Explain the advantages of using digital signals and optical fibres compared with analogue signals and electrical cables for data transmission. *AO1* [4 marks]

Wireless signals

Radio refraction and interference

D–C

- Wireless technology is used by:
  - radio and television
  - laptops
  - mobile phones.
- **Radio waves** are reflected and refracted in the Earth's **atmosphere**:
  - the amount of **refraction** depends on the frequency of the **wave**
  - there is less refraction at higher **frequencies**.
- Radio stations broadcast signals with a particular frequency.
- The same frequency can be used by more than one radio station:
  - the radio stations are too far away from each other to interfere
  - but in unusual weather conditions, the radio waves can travel further and the broadcasts interfere.
- **Interference** is reduced if **digital signals** are used.
- Digital Audio Broadcasting or DAB also provides a greater choice of radio stations but the audio quality is not as good as the FM signals currently used.

B–A*

- DAB eliminates interference between other radio stations.

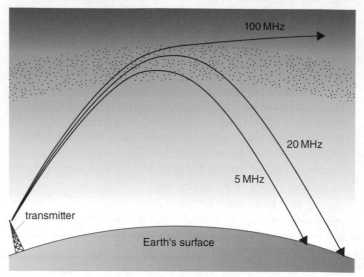

Refraction of waves in the atmosphere

Radio reflection

B–A*

- Radio waves are reflected from the **ionosphere**. They behave like light in an optical fibre and undergo **total internal reflection**.
- Water reflects radio waves but land mass does not.
- Continued reflection by the ionosphere and the oceans allows radio waves to be received from an aerial that is not in line of sight.
- **Microwaves** pass through the ionosphere.
- Microwave signals are received by orbiting **satellites**, amplified and retransmitted back to Earth.
- Communication satellites orbit above the equator and take 24 hours to orbit Earth.

Remember!
Microwave signals are not reflected from satellites.

Communication problems

B–A*

- Radio waves are diffracted when they meet an obstruction.
- Refraction in the atmosphere needs to be taken into account when sending a signal to a satellite.
- The transmitting aerial needs to send a focused beam to the satellite because its aerial is very small.
- The transmitted beam is slightly divergent.
- Some energy is lost from the edge of the transmitting aerial because of **diffraction**.

Improve your grade

Radio communication

The picture shows a transmitter and receiver on the Earth's surface, out of line of sight.

(a) Explain how long wave radio signals travel from the transmitter to the receiver. *AO1* [3 marks]

(b) Explain how microwave signals travel from the transmitter to the receiver. *AO1* [2 marks]

Stable Earth

Earthquake waves

- A seismograph shows the different types of earthquake **wave**.
- L waves travel round the surface very slowly.
- **P waves** are **longitudinal pressure waves**:
  - P waves travel through the Earth at between 5 km/s and 8 km/s
  - P waves can pass through solids and liquids.
- **S waves** are **transverse** waves:
  - S waves travel through the Earth at between 3 km/s and 5.5 km/s
  - S waves can only pass through solids.

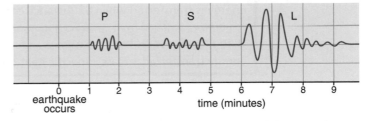

Seismograph trace

D–C

Earth's insides

- P waves travel through the Earth:
  - they are **refracted** by the core
  - the paths taken by P waves mean that scientists can work out the size of the Earth's **core**.
- S waves are not detected on the opposite side of Earth to an earthquake:
  - they will not travel through liquid
  - this tells scientists that Earth's core is liquid.

B–A*

Tan or burn

- A tan is caused by the action of **ultraviolet** light on the skin.
- Cells in the skin produce **melanin**, a pigment that produces a tan.
- People with darker skin do not tan as easily because ultraviolet radiation is filtered out.
- Use a sunscreen with a high SPF, or sun protection factor, to reduce risks.

  maximum length of time to spend in the Sun = published normal burn time × SPF
- People are becoming more aware of the dangers of exposure to ultraviolet radiation, including the use of sun beds.

D–C

Ozone depletion

- At first scientists did not believe there was thinning of the **ozone layer** – they thought their instruments were faulty but other scientists confirmed the results and increased confidence in the findings.

D–C

- **Ozone** is found in the **stratosphere**.
- Ozone helps to filter out ultraviolet radiation.
- **CFC** gases from aerosols and fridges destroy ozone and reduce the thickness of the ozone layer.
  - This increases the potential danger to humans.
- The ozone layer is at its thinnest above the South Pole because ozone depleting chemicals work best in cold conditions.
- Scientists monitor the thickness of the ozone layer using **satellites**.
- There is international agreement to reduce CFC emissions.

B–A*

Improve your grade

Earthquake waves

An Earthquake occurs with its epicentre at **E**. It is detected at two monitoring stations **A** and **B**.

Describe and explain the appearance of the seismograph traces at **A** and **B**.
AO1/AO2 [4 marks]

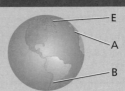

Energy is transferred when a substance changes temperature.

The amount of energy transferred depends on the mass, temperature change and specific heat capacity.

energy transferred = mass × specific heat capacity × temperature change

Heat and temperature

Energy is transferred from a hotter to a colder body.

Temperature is a measure of hotness on an arbitrary scale, measured in °C.

Energy is a measure of energy transfer on an absolute scale measured in J.

Energy is transferred when a substance changes state.

The amount of energy transferred depends on the mass and the specific latent heat.

energy transferred = mass × specific latent heat

Energy transfer

Air is a good insulator and reduces energy transfer by conduction.

Conduction in a solid is by the transfer of kinetic energy.

Trapped air reduces energy transfer by convection.

Convection currents are caused by density changes.

Energy saving in the home can be achieved by:
- double glazing
- cavity wall insulation
- draught strip
- reflecting foil
- loft insulation
- curtains
- careful design.

Shiny surfaces reflect infrared radiation to reduce energy transfer.

Energy transformations can be represented by Sankey diagrams.

$$\text{efficiency} = \frac{\text{useful energy output } (\times 100\%)}{\text{total energy input}}$$

Waves transfer energy

Warm and hot objects emit infrared radiation.

Infrared radiation is used for cooking.

Microwaves can be used for cooking and for communication when transmitter and receiver are in line of sight.

Laser light is single colour and in phase.

Laser light, visible light and infrared are all used to send signals along optical fibres by total internal reflection.

All waves have amplitude, frequency and wavelength

wave speed = frequency × wavelength

Digital and analogue signals are used for communication.

Morse is a digital code.

Digital signals allow many signals to be transmitted at the same time.

Digital signals are clearer.

Radio waves, microwaves, infrared, visible light and ultraviolet are some of the waves in the electromagnetic spectrum.

The energy of the wave increases as the wavelength decreases.

All electromagnetic waves can be reflected, refracted and diffracted.

Radio waves are used for communication.

Longer wavelengths diffract around obstacles.

The stable Earth

Earthquake waves travel through the Earth.

Different waves help us find out about the inside of the Earth.

Exposure to ultraviolet radiation causes sun burn and skin cancer.

Sunscreen and sunblock reduces damage caused by ultraviolet radiation.

CFCs are causing the ozone layer to become thinner.

Collecting energy from the Sun

Photocells

D–C

- The advantages of **photocells** are:
  - they are robust and do not need much maintenance
  - they need no fuel and do not need long power cables
  - they cause no **pollution** and do not contribute to **global warming**
  - they use a **renewable energy** resource.
- The only disadvantage is that they do not produce electricity when it is dark or too cloudy.

How photocells work

B–A*

- A photocell contains two pieces of silicon joined together to make a **p-n junction**.
- One piece has an impurity added to produce an excess of free electrons – n-type silicon.
- The other piece has a different impurity added to produce an absence of free electrons – p-type silicon.
- Sunlight contains energy packets called **photons**.
- Photons cause free electrons to move producing an electric **current**.
- The output from a photocell depends on:
  - the light intensity
  - the surface area exposed
  - the distance from the light source.

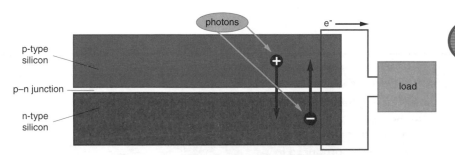

An electric field is set up across the p-n junction

Passive solar heating

B–A*

- The Sun is very hot and produces infrared radiation with a very short **wavelength**:
  - glass is transparent to this short wavelength **radiation**
  - the walls and floor inside a building absorb this radiation, warm up and re-radiate infrared radiation
  - the walls and floor are not as hot as the Sun and the wavelength radiated is therefore longer
  - glass reflects this longer wavelength radiation back inside the building.

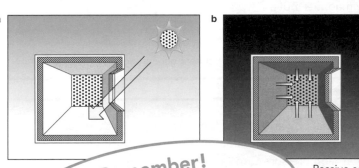

Passive solar heating

Remember!
In the southern hemisphere, the large windows in a house need to face north towards the Sun. In the northern hemisphere they face south.

Heat from the Sun

B–A*

- Solar reflectors are moved by computer to make sure they are always facing the Sun.

Energy from the wind

D–C

- Wind is a renewable form of energy, but it does depend on the speed of the wind:
  - wind **turbines** do not work if there is no wind, nor if the wind speed is too great.
- Wind farms do not contribute to global warming, nor do they **pollute** the **atmosphere** but they can be noisy, take up a lot of space and people sometimes complain that they spoil the view.

Improve your grade

Photocells
A photocell contains two pieces of silicon joined together to make a p-n junction.
Explain how light falling on a photocell produces an electric current. *AO1* [3 marks]

Generating electricity

Bigger currents and voltages

- The current from a **dynamo** can be increased by:
  - using a stronger magnet
  - increasing the number of turns on the coil
  - rotating the magnet faster.
- The output from a dynamo can be displayed on an **oscilloscope**.
- An oscilloscope trace shows how the **current** produced by the dynamo varies with time.
- The time for one complete cycle is called the period of the **alternating current**.

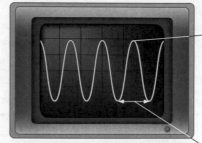

the height of the wave is the maximum (peak) **voltage**

the length of the wave represents the time for one cycle

Oscilloscope trace

Remember!

Frequency = 1 ÷ period
Frequency is measured in **hertz** (Hz)
Period is measured in seconds (s)

Practical generators

- A simple **generator** consists of a coil of wire rotating between the poles of a magnet:
  - the coil cuts through the magnetic field as it spins.
  - a current is produced in the coil.
- A current can be produced if the coil remains stationary and the magnets move.
- Generators at **power stations** work on the same principle.

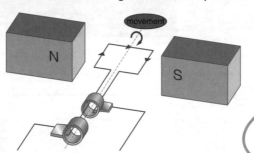

movement

N

S

A simple generating current (AC generator)

Remember!

It is the *relative* movement of magnet and coil that is important.

Power stations

- In conventional power stations, fuels are used to heat water:
  - water boils to produce steam
  - steam at high pressure turns a **turbine**
  - the turbine drives a generator.

Energy efficiency

- **Efficiency** is a measure of how well a device transfers energy.
- Energy in a power station is lost in the boilers, generator and cooling towers.

What is the efficiency of a power station if 60 MJ of fuel energy is converted into 20 MJ of electrical energy?

$$\text{efficiency} = \frac{\text{useful energy output}}{\text{total energy input}}$$

$$= \frac{20\,000\,000}{60\,000\,000}$$

$$= 0.33 \text{ or } 33\%$$

EXAM TIP

Students working at B–A* level should be able to rearrange equations.

EXAM TIP

Always check the wording of your answer is very specific and says exactly what you mean.

Improve your grade

Generators

Describe three ways to increase the output of an electrical dynamo. *AO1* [3 marks]

Global warming

Greenhouse gases

- Most **wavelengths** of electromagnetic **radiation** can pass through the Earth's **atmosphere**, but **infrared** radiation is absorbed.

- Carbon dioxide occurs naturally in the atmosphere as a result of:
  - natural forest fires
  - volcanic eruptions
  - **decay** of dead plant and animal matter
  - its escape from the oceans
  - respiration.

- Man-made **carbon dioxide** is caused by burning fossil fuels, waste incineration, **deforestation** and **cement** manufacture.

- Water vapour is the most significant **greenhouse gas**:
  - almost all of the water vapour occurs naturally
  - a mere 0.001% comes from human activity
  - half of the greenhouse effect is due to water vapour and a further quarter is due to clouds.

- Methane is produced when organic matter decomposes in an environment lacking oxygen.
  - natural sources include wetlands, termites and oceans
  - man-made sources include the mining and burning of **fossil fuels**, the digestive processes in animals such as cattle, rice paddies and the burying of waste in landfills.

The greenhouse effect

- The electromagnetic radiation from the Sun has a relatively short wavelength.

- This radiation is absorbed by and warms the Earth. The Earth then re-radiates the energy as infrared radiation with a longer wavelength.

- This longer wavelength radiation is absorbed by the greenhouse gases which warms the atmosphere.

Dust warms, dust cools

- Dust in the atmosphere can have opposite effects:
  - the smoke from the factories reflects radiation from the town back to Earth. The temperature rises as a result
  - the ash cloud from a volcano reflects radiation from the Sun back into space. The temperature falls as a result.

Scientific data

- It is important that decisions about what to do about **global warming** are made on the basis of scientific evidence, not on the basis of unsubstantiated opinions.

- The vast majority of scientists agree that the evidence supports climate change. The average temperature of the Earth has increased steadily during the past 200 years.

- What scientists do not agree on is the extent to which human activity has contributed.

Remember!
There is disagreement among scientists about the seriousness of global warming. Most scientists agree that the global temperature is rising. They do not agree on more specific elements of the issue: How much will it warm up? What will happen if it does warm up? How far are humans responsible? What should we do to stop it?

EXAM TIP

Students working at B–A* will be expected to interpret data about global warming and climate change.

Improve your grade

The greenhouse effect
Describe the greenhouse effect and explain how it contributes to global warming.
AO1 [4 marks]

Fuels for power

Measuring power

- Electrical appliances usually display a power rating in **watts** (W) or **kilowatts** (kW)

power = voltage × current

> What is the power rating of a kettle working at 9 amps (A) on the mains supply of 230 **volts** (V)?
> power = **voltage** × **current**
> = 9 × 230
> = 2070 W

EXAM TIP

Students should be able to rearrange equations.

energy supplied = power × time

- The unit of electrical **energy** used in the home is the **kilowatt-hour** (kWh).

cost of electricity used = energy used × cost per kWh

> Calculate the cost of using a 250 W television for 30 minutes if one kWh of electricity costs 10p?
> energy used = power × time = 0.25 × 0.5 = 0.125 kWh
> cost of electricity = energy used × cost per kWh = 0.125 × 10 = 1.25p

Cheaper electricity

- We pay less for electricity during the night when not as much is needed, but it still has to be produced.

Energy sources

- Some energy sources are more appropriate than others in a particular situation.
- The choice of energy sources depends on several factors:
  - availability
  - ease of extraction
  - effect on the environment
  - associated risks.

The National Grid

- The **National Grid** is a series of **transformers** and power lines that transport electricity from the power station to the consumer.
- In the National Grid, transformers are used to step up the voltage to as high as 400 000 V. The high voltage leads to:
  - reduced energy loss
  - reduced distribution costs
  - cheaper electricity for consumers.
- Transformers are then used again to step down the voltage to a more suitable level for the consumer.

Transmission losses

- When a current passes through a wire the wire gets hot. The greater the current, the hotter the wire:
  - when a transformer increases the voltage, the current is reduced which means there is less heating effect and therefore less energy lost to the environment.

EXAM TIP

Make sure you give enough detail in your answers. Be guided by the number of marks available for the question.

 Improve your grade

Cost of electricity

Tracey's flat has electric storage heaters which heat up at night and release the heat slowly during the day. Why is this cheaper for Tracey? *AO1* [2 marks]

Nuclear radiations

Ionisation

- **Atoms** contain the same number of **protons** and electrons – this means they are neutral.
- **Ionisation** involves gaining or losing electrons:
  - when the atom gains electrons, it becomes negatively charged
  - when the atom loses electrons, it becomes positively charged.
- The formation of **ions** can cause chemical reactions.
  - Such reactions may disrupt the normal behaviour of molecules inside the body e.g. they may cause strands of **DNA** to break or change; protein molecules may change their shape and these effects are potentially harmful.

Properties of ionising radiations

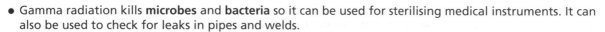

- Alpha, beta and gamma **radiations** come from the **nucleus** of an atom.
- Alpha radiation causes most ionisation and gamma radiation the least.
- Alpha radiation is short ranged (a few centimetres) and is easily absorbed by a sheet of paper or card.
- Beta radiation has a range of about 1 m and is absorbed by a few millimetres of aluminium.
- Gamma radiation is much more penetrating and, although a few centimetres of **lead** will stop most of the radiation, some can pass through several metres of lead or concrete.
- Experiments can be done to identify each type of radiation from its penetrating power - when carrying out these experiments **background radiation** should always be taken into account.

Uses of radioactivity

- Smoke alarms contain a source of alpha radiation:
  - the radiation **ionises** the oxygen and nitrogen atoms in air which causes a very small electric **current** that is detected. When smoke fills the detector in the alarm during a fire, the air is not so ionised, the current is less and the alarm sounds.
- Thicknesses in a paper rolling mill can be controlled using a source of beta radiation and a detector:
  - the amount of radiation passing through the sheet is monitored and the pressure on the rollers adjusted accordingly.
- Gamma radiation kills **microbes** and **bacteria** so it can be used for sterilising medical instruments. It can also be used to check for leaks in pipes and welds.
- The passage of blood and other substances can be traced around the body using a beta or gamma source.

Nuclear waste

- **Plutonium** is a waste product from nuclear reactors which can be used to make nuclear bombs.
- Some low level **radioactive waste** can be buried in landfill sites. High level waste is encased in glass and buried deep underground or reprocessed.
- Radioactive waste can remain radioactive for thousands of years. It must be stored where it cannot leak into natural underground water supplies and hence into lakes and rivers. It is not suitable for making nuclear bombs, but it could be used by terrorists to contaminate water supplies or areas of land.

Advantages and disadvantages of nuclear power

- There are advantages in using nuclear power stations: fossil fuel reserves are not used and no **greenhouse gases** are discharged into the **atmosphere**.
- The disadvantages are its very high maintenance and decommissioning costs, and the risk of accidents similar to the one at Chernobyl.

Improve your grade

Ionising radiation
Give two advantages and two disadvantages of nuclear power. *AO1* [4 marks]

Exploring our Solar System

Our Solar System

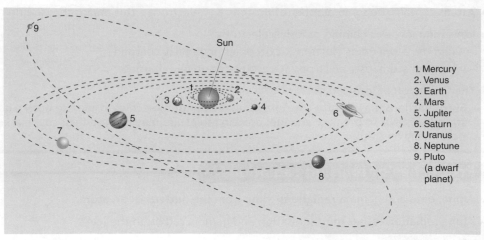

- **Comets** have very **elliptical orbits**:
  - they pass inside the orbit of Mercury and go out well beyond the orbit of Pluto.
- A **meteor** is made from grains of dust that burn up as they pass through the Earth's **atmosphere**:
  - they heat the air around them which glows and the streak is known as a 'shooting **star**'.

1. Mercury
2. Venus
3. Earth
4. Mars
5. Jupiter
6. Saturn
7. Uranus
8. Neptune
9. Pluto (a dwarf planet)

The Solar System

- **Black holes** are formed where large stars used to be:
  - you cannot see a black hole because light cannot escape from it
  - it has a very large mass but a very small size.

- Moons orbit planets and planets orbit stars because **centripetal force** acts on them:
  - centripetal force acts towards the centre of the circular orbit
  - gravitational attraction is the source of the centripetal force.

Exploring the planets

- Unmanned **probes** can go where conditions are deadly for humans.
- Spacecraft carrying humans have to have large amounts of food, water and oxygen aboard.
- Astronauts can wear normal clothing in a pressurised spacecraft.
- Outside the spacecraft they need to wear special spacesuits:
  - a dark visor stops an astronaut being blinded
  - the suit is pressurised and has an oxygen supply for breathing
  - the surface of the suit facing towards the Sun can reach 120 °C
  - the surface of the suit facing away from the Sun may be as cold as −160 °C
- When travelling in space, astronauts are subjected to lower gravitational forces than on Earth.
- Unmanned spacecraft cost less and do not put astronauts lives at risk:
  - they have to be very reliable because there is usually no way of repairing them when they break down.

A long way to go!

- Distances in space are very large.
- Light travels at 300 000 km/s:
  - light from the Sun takes about eight minutes to reach us on Earth
  - light from the next nearest star (Proxima Centauri) takes 4.22 years.
- America plans a manned mission to Mars after 2020 at a cost of £400 billion.

Remember!
A **light-year** is the distance light travels in one year.

EXAM TIP

Sometimes a word is emphasised in the question to draw your attention to its importance. Be careful to answer accordingly.

Improve your grade

Unmanned space travel

Why do we send *unmanned* space probes to explore our Solar System? *AO1* [3 marks]

Threats to Earth

Asteroids

- **Asteroids** are mini-planets or planetoids orbiting the Sun:
  - most orbit between Mars and Jupiter
  - they are large rocks that were left over from the formation of the **Solar System**.

Asteroid or planet

- All bodies in space, including planets, were formed when clouds of gas and dust collapsed together due to gravitational forces of attraction.
- The mass of an object determines its gravitational force.
- Asteroids have relatively low masses compared to the mass of Jupiter.
- Jupiter's gravitational force prevents asteroids from joining together to form another planet.

Origin of the Moon

- Scientists believe our Moon was a result of the collision between two planets in the same orbit. The iron **core** of the other planet melted and joined with the Earth's core, less dense rocks began to orbit and they joined together to form our Moon.

- There is scientific evidence which supports this idea.
  - The average **density** of Earth is 5500 kg/m³ while that of the Moon is only 3300 kg/m³.
  - There is no iron in the Moon.
  - The Moon has exactly the same oxygen composition as the Earth, but rocks on Mars and meteorites from other parts of the Solar System have different oxygen compositions.

Evidence for asteroids

- Geologists examine evidence to support the theory that asteroids have collided with Earth:
  - near to a crater thought to have resulted from an asteroid impact, they found quantities of the metal iridium – a metal not normally found in the Earth's **crust** but common in meteorites
  - many fossils are found below the layer of iridium, but few fossils are found above it
  - **tsunamis** have disturbed the fossil layers, carrying some fossil fragments up to 300 km inland.

A comet's orbit

- Most **comets** pass inside the orbit of Mercury and well beyond the orbit of Pluto:
  - as the comet passes close to the Sun, the ice melts and solar winds blow the dust into the comet's tail which always points away from the Sun.
- Scientists are constantly monitoring and plotting the paths of comets and other **near-Earth objects** NEOs.

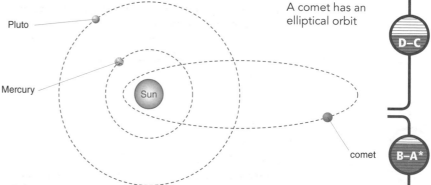

A comet has an elliptical orbit

- The speed of a comet increases as it approaches the Sun and decreases as it gets further away. This is because of the changing gravitational attraction.

NEOs

- If an NEO is on a collision course with Earth, it could be the end of life on Earth. To avoid this, one option would be to explode a rocket near to the NEO which could alter its course enough to miss Earth.

Improve your grade

Asteroids

What evidence have scientists gathered to show how the Moon was formed?
AO1 [3 marks]

The Big Bang

The expanding Universe

- Almost all of the galaxies are moving away from each other with the further galaxies moving fastest.

Models of the Universe

- With the help of the newly invented telescope, Galileo observed four moons orbiting Jupiter. This confirmed that not everything orbited the Earth and supported Copernicus' idea that planets orbit the Sun.

- The Roman Catholic Church did not support Galileo's model as they believed that the Earth was at the centre of the **Universe** and it was a very long time before it was accepted.

- In the 17th century, Newton was working on his theory of universal gravitation which suggested that all bodies **attract** one another.

- Today, we believe that gravitational collapse is prevented because the Universe is constantly expanding as a result of the **Big Bang**.

Red shift

- When a source of light is moving away from an observer, its **wavelength** appears to increase which shifts light towards the red end of the spectrum – **red shift**.

- When scientists look at light from the Sun, there is a pattern of lines across the spectrum. This same pattern is observed when they look at light from distant **stars** but it is closer to the red end of the spectrum.

- Scientists can use information from red shift to work out the age of the Universe.

A star's life history

- The swirling cloud of gas and dust is a nebula:
  - nebula clouds are pulled together by **gravity** and, as the spinning ball of gas starts to get hot, it glows. This protostar cannot be seen because of the dust cloud
  - gravity causes the star to become smaller, hotter and brighter and after millions of years, the **core** temperature is hot enough for nuclear **fusion** to take place. As hydrogen nuclei join together to form **helium** nuclei, energy is released and the star continues to shine while there is enough hydrogen.

- Small stars shine for longer than large stars because they have less hydrogen but use it up at a slower rate and what happens at the end of a star's life depends on its size.

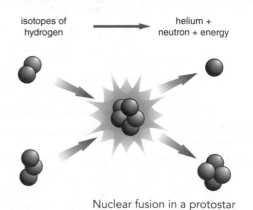

isotopes of hydrogen → helium + neutron + energy

Nuclear fusion in a protostar

The end of a star's life

- A medium-sized star, like the Sun, becomes a red giant: while the core contracts, the outer part cools, changes colour from yellow to red and expands:
  - gas shells, called planetary nebula, are thrown out
  - the core becomes a white dwarf shining very brightly but eventually cools to become a black dwarf

- Large stars become red supergiants: as the core contracts and the outer part expands and, suddenly, the core collapses to form a neutron star and there is an **explosion** called a supernova
  - neutron stars are very dense
  - remnants from a supernova can merge to form a new star
  - the core of the neutron star continues to collapse, becomes even more dense and could form a **black hole**.

- a black hole has a very large mass concentrated in a small volume so it has a very large **density** and its large mass means it has a very strong gravitational pull.

Improve your grade

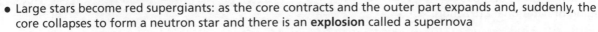

The Big Bang

What is red shift and how does it provide evidence for the Big Bang? *AO1* [4 marks]

P2 Summary

Kinetic energy from moving air turns the blades on a wind turbine to produce electricity.

Passive solar heating uses glass to help keep buildings warm.

Energy from the Sun

Some gases in the Earth's atmosphere trap heat from the Sun increasing global warming.

Scientists disagree about how much effect humans have on the increase in levels of these greenhouse gases.

Photocells do not need fuel or cables, need little maintenance and cause no pollution.

A dynamo produces electricity when coils of wire rotate inside a magnetic field.

The size of the current depends on the number of turns, the strength of the field and the speed of rotation.

In power stations, fuels release energy as heat.

Water is heated to produce steam.

The steam drives turbines.

Turbines turn generators.

Generators produce electricity.

$$\text{efficiency} = \frac{\text{energy output}}{\text{energy input}}$$

Electricity generation

Transformers change the size of the voltage and current.

The National Grid transmits electricity around the country at high voltage and low current.

This reduces energy loss.

Nuclear fuels are radioactive.

The radiation produced can cause cancer.

Waste products remain radioactive for a long time.

The main forms of ionising radiation are alpha, beta and gamma.

Their uses depend on their penetrative and ionisation properties.

Planets, asteroids and comets orbit the Sun in our Solar System.

Centripetal forces keep bodies in orbit.

Medium-sized stars, such as our Sun, were formed from nebulae and will eventually become red giants, white dwarfs and finally black dwarfs.

Our Solar System

When two planets collide, a new planet and a moon may be formed.

The Universe consists of many galaxies.

Models of the Universe have changed over time and sometimes these changes take a long time to be accepted.

The Universe is explored by telescopes on the Earth and in space.

Large distances mean that it takes a long time for information to be received and inter-galactic travel unlikely.

The Universe

Most asteroids orbit between Mars and Jupiter but some pass closer to the Earth. They are constantly being monitored. An asteroid strike could cause climate change and species extinction.

Scientists believe that the Universe started with the Big Bang.

The evidence is red shift.

Speed

Measuring speed

- The formula for **speed** is:

$$\text{average speed} = \frac{\text{distance}}{\text{time}}$$

- We write '**average speed**' because the speed of a car changes during a journey.

> An aircraft travels 1800 km in 2 hours.
> Average speed $= \frac{1800}{2} = 900$ km/h $= 900 \times \frac{1000}{3600} = 250$ m/s

Remember!
There are 3600 seconds in one hour.

- The speed of a car at a certain point in time is called **instantaneous speed**.

distance travelled = average speed × time
$$= \frac{(u + v)t}{2}$$

Remember!
You may be asked to rearrange equations. Practise doing this.

- Increasing the **speed** means increasing the distance travelled in the same time. Increasing the speed reduces the time needed to cover the same distance.

Distance–time graphs

- **Distance–time graphs** allow a collection of data to be shown. It is easier to interpret data when they are plotted on a graph than when they are listed in a results table.

- The **gradient** of a distance–time graph tells you about the speed of the object. A higher speed means a steeper gradient.

- In graph **a**, the distance travelled by the object each second is the same. The gradient is constant, so the speed is constant. In graph **b**, the distance travelled by the object each second increases as the time increases. The gradient increases, so there is an increase in the speed of the object.

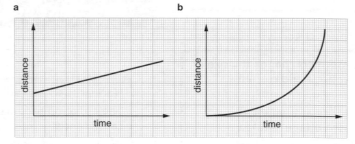

Distance–time graphs

- Speed is equal to the gradient of a distance–time graph; the higher the speed the steeper the gradient.

- A **straight line** indicates that the speed is constant.

- A **curved line** shows that the speed is changing.

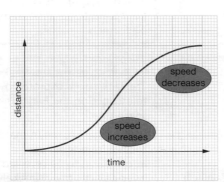

A curved line shows that the speed is not constant

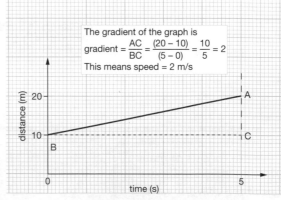

The gradient of the graph is
$$\text{gradient} = \frac{AC}{BC} = \frac{(20 - 10)}{(5 - 0)} = \frac{10}{5} = 2$$
This means speed = 2 m/s

The higher the speed, the steeper the gradient

Improve your grade

Calculating time

How many hours will it take to travel 560 km at an average speed of 25 m/s?
AO1 [1 mark] *AO2* [2 marks]

Changing speed

Speed–time graphs

- A change of **speed** per unit time is called **acceleration**.
- If the speed is increasing, the object is accelerating. If the speed is decreasing, the object is decelerating.
- The area under a speed–time graph is equal to the **distance** travelled.
  - The speed of car B in the graph is increasing more rapidly than the speed of car A, so car B is travelling further than car A in the same time.
  - The area under line B is greater than the area under line A for the same time.
  - The speed of car D is decreasing more rapidly than the speed of car C, so car D isn't travelling as far as car C in the same time.
  - The area under line D is smaller than the area under line C for the same time.

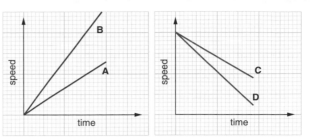

Speed–time graphs for 4 cars – A, B, C and D

EXAM TIP

Don't confuse distance–time and speed–time graphs. Always look at the axes carefully.

 D–C

Acceleration

- The formula for measuring acceleration is:

$$\text{acceleration} = \frac{\text{change in speed (or velocity)}}{\text{time taken}}$$

- Acceleration is measured in metres per second squared (m/s^2).
- A negative acceleration shows the car is decelerating.

> A new car boasts a rapid acceleration of 0 to 108 km/h in 6 seconds.
> A speed of 108 km/h is $\frac{108 \times 1000}{60 \times 60} = 30$ m/s
>
> $\text{Acceleration} = \frac{\text{change in speed}}{\text{time taken}} = \frac{(30 - 0)}{6} = 5$ m/s^2
>
> This means the speed of the car increases by 5 m/s every second.

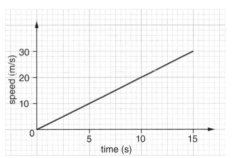

Acceleration is the gradient of a speed–time graph.

 D–C

Relative velocities

- **Velocity** is a vector – it has both size (speed) and direction.
- When two cars are moving past each other, their **relative velocity** is:
  - the sum of their individual velocities if they are going in opposite directions
  - the difference of their individual velocities if they are going in the same direction.

D–C

Circular motion

- A vehicle may go around a roundabout at a constant speed but it is accelerating. This is because its direction of travel is changing; it's not going in a **straight line**. The driver needs to apply a **force** towards the centre of the roundabout to change direction. This gives the vehicle an acceleration directed towards the centre of the roundabout.
- Any object moving along a circular path moves at a tangent to the circle, or arc of a circle.

B–A*

Improve your grade

Calculating acceleration

A car is travelling at 10 m/s. It accelerates at 4 m/s^2 for 8 s. How fast is it then going?
AO1 [1 mark] *AO2* [2 marks]

Forces and motion

Force, mass and acceleration

- If the **forces** acting on an object are balanced, it's at rest or has a constant speed. If the forces acting on an object are unbalanced, it speeds up or slows down.

- The unit of force is the **newton** (N).

  force = mass × **acceleration**

  Where F = **unbalanced** force in N, m = mass in kg and a = acceleration in m/s².

  > Marie pulls a sledge of mass 5 kg with an acceleration of 2 m/s² in the snow.
  > The force needed to do this is: F = ma = 5 × 2 = 10 N

- The equation F = ma is used to find mass or acceleration if the **resultant force** is known.

  > Professional golfers hit a golf ball with a force of approximately 9000 N. If the mass of the ball is 45 g, the acceleration during the very short time (about 0.005 milliseconds) of impact can be calculated.
  >
  > $a = \dfrac{F}{m} = \dfrac{9000}{0.045} = 200\,000$ m/s²

Car safety

- **Reaction time**, and therefore thinking distance, may increase if a driver is:
  - tired
  - under the influence of alcohol or other drugs
  - travelling faster
  - distracted or lacks concentration.

- **Braking distance** may increase if:
  - the road conditions are poor e.g. icy
  - the car has not been properly maintained e.g. worn brakes
  - the speed is increased.

- For safe driving, it is important to be able to stop safely:
  - Keep an appropriate distance from the car in front.
  - Have different speed limits for different types of road and locations.
  - Slow down when road conditions are poor.

- Factors affecting braking distance are:
  - The greater the mass of a vehicle the greater its braking distance.
  - The greater the speed of a vehicle the greater its braking distance.
  - When the brakes are applied the brake pads are pushed against the disc. This creates a large **friction** force that slows the car down. Worn brakes reduce the friction force, increasing the braking distance.
  - Worn tyres with very little **tread** reduce the grip of the wheels on a slippery road, leading to skidding and an increase in braking distance.
  - Increased braking force reduces the **stopping distance**.

- **Thinking distance** increases **linearly** with speed.

- Braking distance increases as a squared relationship e.g. the braking distance at 60 mph is nine times the braking distance at 20 mph.

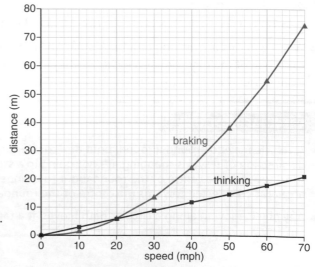

Thinking distance and speed is proportional, as is braking distance and speed

Improve your grade

Stopping distance

Explain why brakes and tyres are checked when a car has its annual MOT test. *AO2* [2 marks]

Work and power

Work

- **Work** is done when a **force** moves an object in the direction in which the force acts.
- The formula for **work done** is:

  work done = force × distance moved (in the direction of the force)

  > If a person weighs 700 N, the work he does against **gravity** when he jumps 0.8 m is:
  > work done = force × distance moved = 700 × 0.8 = 560 J

Weight

- **Weight** is a measure of the gravitational attraction on a body acting towards the centre of the Earth.
- The formula for weight is:

  weight = mass × **gravitational field strength**

- A mass of 1 kg has a weight of about 10 N on Earth.

Power

- **Power** is the rate at which work is done.
- The formula for power is:

  $$\text{power} = \frac{\text{work done}}{\text{time taken}}$$

- The equation for power can be rearranged to work out:

  work done = power × time

  $$\text{time taken} = \frac{\text{work done}}{\text{power}}$$

  > When the Eurostar travels at maximum **speed**, its power is 2 MW. The amount of work done, or energy transferred, in 2 hours is calculated by:
  >
  > work done = power × time = 2 000 000 × (2 × 60 × 60) = 14 400 000 000 J
  >
  > At maximum power, 14 400 MJ of energy would be transferred to other forms during a 2 hour journey. The Eurostar operates at maximum power for only a short part of the journey.

- A person's power is greater when they run than when they walk.
- Some cars are more powerful than others. They travel faster and cover the same distance in a shorter time and require more fuel. The power rating of a car depends on its engine size. More powerful cars have greater **fuel consumption**.
- Fuel is expensive and a car with high fuel consumption is expensive to run.
- Fuel **pollutes** the environment.
  - Car **exhaust gases**, especially **carbon dioxide**, are harmful.
  - Carbon dioxide is also a major source of **greenhouse gases**, which contribute to climate change.
- Power is also related to force and speed.

  $$\text{power} = \frac{\text{work done}}{\text{time taken}} = \frac{\text{force} \times \text{distance}}{\text{time taken}} = \text{force} \times \frac{\text{distance}}{\text{time}} = \text{force} \times \text{speed}$$

Improve your grade

Calculating driving force

A Ford Focus has a power rating of 104 kW.

(a) Calculate the resultant force acting when the car is travelling at 90 km/h.

(b) Explain how this force compares with the driving force of the engine.

   AO1 [3 marks] *AO2* [1 mark]

Energy on the move

Kinetic energy

- The **braking distance** of a car increases with increasing **speed**, but not proportionally.
- The formula for **kinetic energy** is:

  kinetic energy = $\frac{1}{2}$ mv^2, where m = mass in kg, v = **velocity** in m/s

  > If a car has a mass of 1000 kg, its kinetic energy:
  > – at 20 m/s is $\frac{1}{2}$ mv^2 = $\frac{1}{2}$ × 1000 × (20)^2 = 200 000 J
  > – at 40 m/s is $\frac{1}{2}$ mv^2 = $\frac{1}{2}$ × 1000 × (40)^2 = 800 000 J

- When a car stops, its kinetic energy changes into heat in the brakes, tyres and road.
- This can be shown by the formula:

  work done by brakes = loss in kinetic energy
- The change in kinetic energy can be shown in the formula:

  braking force × braking distance = loss in kinetic energy
- When the speed of the car doubles, the kinetic energy and the braking distance quadruple.
  – This is why there are speed limits and penalties for drivers who exceed them.

Fuel

- **Fuel consumption** data are based on ideal road conditions for a car driven at a steady speed in urban and non-urban conditions.

car	fuel	engine in litres	miles per gallon (mpg)	
			urban	non-urban
Renault Megane	**petrol**	2.0	25	32
Land Rover	petrol	4.2	14	24

> ### EXAM TIP
> Make sure you can interpret fuel consumption tables.

- Factors that affect the fuel consumption of a car are:
  – the amount of **energy** required to increase its kinetic energy
  – the amount of energy required for it to do work against **friction**
  – its speed
  – how it is driven, such as excessive **acceleration** and **deceleration**, constant braking and speed changes
  – road conditions, such as a rough surface.

Electrically powered cars

- Electric cars are **battery** driven or **solar-powered**.
  – The battery takes up a lot of room.
  – They have limited range before **recharging**.
  – They are expensive to buy but the cost of recharging is low.
  – Solar-powered cars rely on the Sun shining and need backup batteries.
- Exhaust fumes from petrol-fuelled and diesel-fuelled cars cause serious **pollution** in towns and cities.
- Battery-driven cars do not **pollute** the local environment, but their batteries need to be recharged. Recharging uses electricity from a **power station**. Power stations pollute the local atmosphere and cause acid rain.
- Solar-powered cars do not produce any **carbon dioxide** emissions.
- **Biofuels** may reduce carbon dioxide emissions but this is not certain because **deforestation** leads to an increase in carbon dioxide levels.

Improve your grade

Carbon dioxide emissions

Some scientists suggest that carbon dioxide emissions from burning biofuels may be at least 20% lower than those from fossil fuels. Some scientists argue that overall the emissions may be higher than from fossil fuels. Suggest why emissions may be higher. *AO3* [3 marks]

Crumple zones

Momentum and force

- The formula for **momentum** is:

  momentum = mass × velocity and the units are kgm/s

- To reduce injuries in a collision, **forces** should be as small as possible.

  $$force = \frac{change\ in\ momentum}{time}$$

- Spreading the momentum change over a longer time reduces the force.

- To minimise injury, forces acting on the people in a car during an accident must be minimised.

  force = mass × **acceleration**

- Force can be reduced by reducing the acceleration. This is done by:
  - increasing stopping or collision time
  - increasing stopping or collision distance.

- Safety features that do this include:
  - crumple zones – seat belts – **air bags** – crash barriers – **escape lanes**.

Car safety features

- Modern cars have safety features that absorb **energy** when a vehicle stops suddenly. These are:
  - brakes that get hot
  - **crumple zones** that change shape
  - **seat belts** that stretch a little
  - air bags that inflate and squash.

- On impact:
  - Crumple zones at the front and rear of the car absorb some of its energy by changing shape or 'crumpling'.
  - Seat belts stretch a little so that some of the person's **kinetic energy** is converted to elastic energy.
  - Air bags absorb some of the person's kinetic energy by squashing up around them.

- All these safety features:
  - absorb energy
  - change shape
  - reduce injuries
  - reduce momentum to zero slowly, therefore reducing the force on the occupants.

- Some people do not like wearing seat belts because:
  - there is a risk of chest injury
  - they may be trapped in a fire
  - drivers may be encouraged to drive less carefully because they know they have the protection from a seat belt.

- Despite **computer modelling**, crash tests using real vehicles and **dummies** provide more safety information.

- **ABS brakes** are a **primary safety feature** which helps to prevent a crash.

- In the ABS system, wheel-speed **sensors** send information to a computer about the rotational **speed** of the wheels. The computer controls the pressure to the brakes, via a pump, to prevent the wheels locking up. This increases the braking force (F) just before the wheels start to skid.

  kinetic energy lost = **work done** by the brakes

  - $\frac{1}{2}$ mv² = Fd where m = mass of car, v = speed of car before braking, d = **braking distance**.
  - If F increases, the braking distance (d) decreases.

- Other primary safety features include:
  - **cruise control** which stops a driver becoming tired on a long journey and pressing harder on the accelerator
  - electric windows and **paddle shift controls** which allow the driver to concentrate on the road.

Improve your grade

Seat belts

Some people think that wearing a seat belt should be up to the individual, not the law. Explain how the wearing of seat belts can help to avoid injury but may not always do so. *AO1* [3 marks]

Falling safely

Falling objects

- All objects fall with the same **acceleration** due to **gravity** as long as the effect of **air resistance** is very small.

- The size of the air resistance **force** on a falling object depends on:
  - its **cross-sectional area** – the larger the area the greater the air resistance
  - its **speed** – the faster it falls the greater the air resistance.

- Air resistance has a significant effect on motion only when it is large compared to the **weight** of the falling object.

- The speed of a **free-fall** parachutist changes as he falls to Earth.

1   2   3   4

600 N
600 N
600 N
600 N

1000 N
600 N

600 N
600 N

- In picture 1, the weight of the parachutist is greater than air resistance. He accelerates.

- In picture 2, the weight of the parachutist and air resistance are equal. The parachutist has reached **terminal speed** because the forces acting on him are balanced.

- In picture 3, the air resistance is larger than the weight of the parachutist. He slows down and air resistance decreases.

- In picture 4, the air resistance and weight of the parachutist are the same. He reaches a new, slower terminal speed.

Terminal speed

- In picture 1, the parachutist accelerates, displacing more air molecules every second. The air resistance force increases. This reduces his acceleration. So, the higher the speed, the more air resistance.

- In picture 2, the parachutist's weight is equal to the air resistance; the forces on him are balanced. He travels at a constant speed – terminal speed.

- In picture 3, when the parachute opens, the upward force increases suddenly as there is a much larger surface area, displacing more air molecules every second. So, the larger the area, the more air resistance. The parachutist decelerates, displacing fewer air molecules each second, so the air resistance force decreases.

- In picture 4, the parachutist reaches a new slower terminal speed when his weight is equal to the air resistance once more, so he lands safely.

- **Drag** racers and the Space Shuttle use parachutes to slow them down rapidly.

Gravitational field strength – g

- The force on each kilogram of mass due to gravity, g = 10 N/kg approximately on Earth.
  - g is also known as acceleration due to gravity, g = 10 m/s^2 approximately.

- The **gravitational field strength**:
  - is unaffected by atmospheric conditions
  - varies with position on the Earth's surface (9.78 N/kg at the equator and 9.83 N/kg at the poles)
  - varies with height above or depth below the Earth's surface.

Improve your grade

Gravitational field strength

(a) Explain why a 1 kg ball dropped from a height of 10 m above the ground at the North Pole, takes less time to reach the ground than an identical 1 kg ball dropped from 10 m above the ground at the equator.

(b) How would the time taken be different if the ball was taken to the top of a mountain and dropped from a height of 10 m?
AO1 [2 marks] AO2 [1 mark]

The energy of games and theme rides

Gravitational potential energy

- An object held above the ground has **gravitational potential energy**.
- This is shown by the formula:

  GPE = mgh where

  m = mass, h = vertical height moved,
  g = **gravitational field strength** (10 N/kg)

- GPE is measured in **joules** (J).

Remember!
You will be expected to rearrange equations. Practise doing this.

D–C

Energy transfers

- A bouncing ball converts gravitational potential energy to **kinetic energy** and back to gravitational potential energy. It does not return to its original height because **energy** is transferred.

B the ball has *just* reached the ground so it has kinetic energy but no gravitational potential energy

D the ball has gravitational potential energy but no kinetic energy at the top of the bounce

2.00 m

1.20 m

A   B   C   D

A the ball has gravitational potential energy but no kinetic energy

C the ball squashes and has **elastic potential energy** which is converted to kinetic energy as it leaves the ground

Stages of energy transfer when a ball is dropped

D–C

- When skydivers reach **terminal speed**, their kinetic energy ($\frac{1}{2}mv^2$) has a maximum value and remains constant. The gravitational potential energy lost as they fall is used to do **work** against **friction (air resistance)**.
  - When terminal speed is reached it can be shown as:

    change in GPE = work done against friction

B–A*

How a roller coaster works

- A roller coaster uses a motor to haul a train up in the air. The riders at the top of a roller coaster ride have a lot of gravitational potential energy.

- When the train is released it converts gravitational potential energy to kinetic energy as it falls. This is shown by the formula:

  loss of gravitational potential energy (GPE) = gain in kinetic energy (KE)

D–C

- Ignoring friction, as the train falls:

  $mgh = \frac{1}{2}mv^2$

  $h = \dfrac{mv^2}{2\,mg} = \dfrac{v^2}{2\,g}$

  - so this is independent of the mass of the falling object.

B–A*

- Each peak is lower than the one before because some energy is transferred to heat and sound due to friction and air resistance.

- This is shown by the formula:

  GPE at top = KE at bottom + energy transferred (to heat and sound) due to friction

- If **speed** doubles, KE quadruples (KE $\propto v^2$).

- If mass doubles, KE doubles (KE $\propto m$).

D–C

Improve your grade

Energy changes during free-fall
Mel is a free-fall parachutist. During her time in free fall she reaches terminal speed.
Explain how her gravitational potential energy and kinetic energy change during her descent.
AO1 [3 marks]

P3 Summary

Speed and acceleration

The gradient of a distance–time graph is the speed.

The steeper the gradient, the higher the speed.

$$\text{distance} = \text{average speed} \times \text{time}$$

$$\text{acceleration} = \frac{\text{change in speed}}{\text{time}}$$

Relative velocity is the difference between the velocities of two bodies having regard to their direction.

The area under a speed–time graph is the distance travelled.

The gradient of a speed–time graph is the acceleration.

The steeper the gradient, the greater the acceleration.

A negative gradient means the body is slowing down.

Forces and motion

Forces always occur in pairs:
- that are the same size
- act in opposite directions
- act on different objects.

Forces can make things speed up or slow down.

$$\text{force} = \text{mass} \times \text{acceleration}$$

$$\text{weight} = \text{mass} \times \frac{\text{gravitational}}{\text{field strength}}$$

When a car stops the

$$\frac{\text{total}}{\text{stopping}}_{\text{distance}} = \frac{\text{thinking}}{\text{distance}} + \frac{\text{braking}}{\text{distance}}$$

Thinking distance depends on the state of the driver and the speed of the vehicle (linear).

Braking distance depends on the state of the road, vehicle and speed (squared).

Work, energy and power

Moving objects possess kinetic energy.

The faster they travel, the more kinetic energy they possess.

The greater their mass, the more KE they possess.

$$KE = \tfrac{1}{2}mv^2$$

Fuel consumption depends on the energy required to increase kinetic energy.

Roller coasters use gravitational potential energy as the source of movement.

GPE = mgh

During a roller coaster ride, GPE is transferred into KE and back again.

Work (J) is done when a force moves through a distance.

Energy (J) is needed to do work.

Power (W) is a measure of how quickly work is done.

$$\text{work done} = \text{force} \times \text{distance}$$

$$\text{power} = \frac{\text{work done}}{\text{time}}$$

$$\text{power} = \text{force} \times \text{speed}$$

Modern cars have lots of safety features that absorb energy when the cars stop:
- crumple zones
- seat belts
- air bags.

$$\text{force} = \frac{\text{change in momentum}}{\text{time}}$$

Spreading momentum change over a longer time reduces injury.

ABS brakes mean the driver can keep control when braking without skidding.

Falling safely

Terminal speed is the maximum speed reached by a falling object. This happens when the forces acting on the object are balanced.

Falling objects get faster as they fall.

They are pulled towards the centre of the Earth by their weight.

The acceleration of all objects is the same at any point on the Earth's surface.

Gravitational field strength varies across the Earth's surface and with height above the surface.

The forces on a falling object depend on:
- the speed of the object (higher speed means more drag)
- the area of the object (larger area means more drag).

Sparks

Electrons

- An **atom** consists of a small positively charged **nucleus** surrounded by an equal number of negatively charged **electrons**.
- In a stable, neutral atom there are the same amounts of positive and negative **charges**.
- All electrostatic effects are due to the movement of electrons.
- The law of electric charge states that: like charges **repel**, unlike charges **attract**

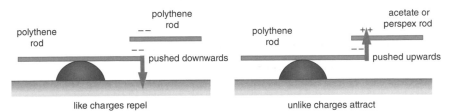

Like charges repel and unlike charges attract

D–C

- When a polythene rod is rubbed with a duster, electrons are transferred from the duster to the polythene, making the polythene rod negatively charged.
- When an acetate rod is rubbed with a duster, electrons are transferred from the acetate to the duster, leaving the acetate rod positively charged.
- In general an object has:
  - a negative charge due to an excess of electrons
  - a positive charge due to a lack of electrons.
- atoms or molecules that have become charged are ions.

Remember!
Only the electrons can be transferred.

B–A*

Electrostatic shocks

- When inflammable gases or vapours are present, or there is a high concentration of oxygen, a **spark** from **static** electricity could ignite the gases or vapours and cause an **explosion**.
- If a person touches something at a high **voltage**, large amounts of electric charge may flow through their body to earth.
- Even small amounts of charge flowing through the body can be fatal.
- Static electricity can be a nuisance but not dangerous.
  - Dust and dirt are attracted to insulators, such as television screens.
  - Clothes made from synthetic materials often 'cling' to each other and to the body.

D–C

- Electric **shocks** can be avoided in the following ways:
  - if an object that is likely to become charged is connected to earth, any build-up of charge would immediately flow down the **earth wire**
  - in a factory where machinery is at risk of becoming charged, the operator stands on an insulating rubber mat so that charge cannot flow through them to earth
  - shoes with insulating soles are worn by workers if there is a risk of charge building up so that charge cannot flow through them to earth
  - fuel tankers are connected to an aircraft by a conducting cable during refuelling.
- Anti-static sprays, liquids and cloths made from conducting materials carry away electric charge. This prevents a build-up of charge.

B–A*

 Improve your grade

Static charge

Connor is in the library walking on a nylon carpet. He touches a metal bookshelf and receives an electric shock. Explain how he became charged and why he received a shock.
AO2 [3 marks]

Uses of electrostatics

Dust precipitators

- A dust precipitator removes harmful particles from the chimneys of factories and power stations that **pollute** the atmosphere.

- A metal grid (or wires) is placed in the chimney and given a large **charge** from a high-**voltage** supply.

- Plates inside the chimney are **earthed** and gain the opposite charge to the grid.

- As the dust particles pass close to the grid, they become charged with the same charge as the grid.

- Like charges **repel**, so the dust particles are repelled away from the wires. They are **attracted** to the oppositely charged plates and stick to them.

- At intervals the plates are vibrated and the dust falls down to a collector.

- The dust particles gain or lose **electrons** to become charged.

- The charge on the dust particles induces a charge on the earthed metal plate.

- Opposite charges attract so the dust is attracted to the plate.

Paint spraying

- Static electricity is used in paint spraying.
  - The spray gun is charged.
  - All the paint particles become charged with the same charge.
  - Like charges repel, so the paint particles spread out giving a fine spray.
  - The object to be painted is given the opposite charge to the paint.
  - Opposite charges attract, so the paint is attracted to the object and sticks to it.
  - The object gets an even coat, with limited paint wasted.

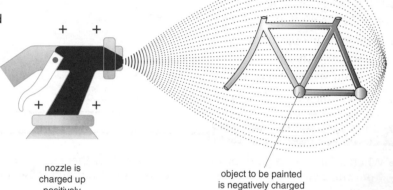

nozzle is charged up positively

object to be painted is negatively charged

Electrostatic principles in paint spraying

- If the object to be painted is not charged, the paint moves onto it but:
  - the object becomes charged from the paint, gaining the same charge
  - further paint droplets are repelled away from the object.

- Therefore the object to be painted is given the opposite charge to the paint. If the paint is negatively charged, having gained electrons, the object should be positively charged, by losing electrons.

Defibrillators

- **Defibrillation** is a procedure to restore a regular heart rhythm by delivering an electric **shock** through the chest wall to the heart.
  - Two **paddles** are charged from a high-voltage supply.
  - They are then placed firmly on the patient's chest to ensure good electrical contact.
  - Electric charge is passed through the patient to make their heart contract.
  - Great care is taken to ensure that the operator does not receive an electric shock.

- If a defibrillator is switched on for 5 milliseconds (0.005 s), the power can be calculated from:

$$\text{power} = \frac{\text{energy}}{\text{time}} = \frac{400}{0.005} = 80\,000 \text{ W}$$

Improve your grade

Spray painting

Static electricity is useful in spray-painting cars.

Explain how by writing about:
- electrostatic charge
- electrostatic force
- why it is used. *AO1* [3 marks]

Safe electricals

Resistance

- A **variable resistor**, or **rheostat**, changes the **resistance**. Longer lengths of wire have more resistance; thinner wires have more resistance.

- **Voltage (potential difference)** is measured in **volts (V)** using a **voltmeter** connected in parallel.
  - For a fixed resistor, as the voltage across it increases, the **current** increases.
  - For a fixed power supply, as the resistance increases, the current decreases.

- The formula for resistance is:

$$\text{resistance} = \frac{\text{voltage}}{\text{current}} \qquad R = \frac{V}{I}$$

- Resistance is measured in **ohms (Ω)**.

- The formula for resistance can be rearranged to find out:

$$\text{voltage } V = IR \qquad \text{or} \qquad \text{current } I = \frac{V}{R}$$

> **Remember!**
> Always remember to include the correct unit with your answer.

Live, neutral and earth wires

- The **live wire** carries a high voltage around the house.

- The **neutral wire** completes the circuit, providing a return path for the current.

- The **earth wire** is connected to the case of an appliance to prevent it becoming live.

- A **fuse** contains wire which melts, breaking the circuit, if the current becomes too large.

- No current can flow, preventing overheating and further damage to the appliance.

- Earth wires and fuses stop a person receiving an electric **shock** if they touch a faulty appliance. As soon as the case becomes 'live', a large current flows in the earth and live wires and fuse 'blows'.

- A re-settable fuse (**circuit-breaker**) doesn't need to be replaced to restore **power**; it can be re-set.

> **Remember!**
> Double insulated appliances do not need an earth wire as the outer case is not a conductor.

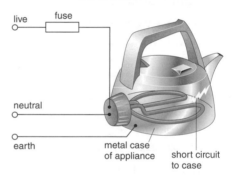

The arrangement of wires in metal-cased appliance

Electrical power

- The rate at which an appliance transfers energy is its power rating:

  power = voltage × current

- The formula for electrical power can be used to calculate the correct fuse to use in an electrical device, e.g.

power	= voltage × current
current	$= \dfrac{\text{power}}{\text{voltage}}$
mains voltage	= 230 V
power of kettle	= 2500 W
current	$= \dfrac{2500}{230} = 10.9$ A

Therefore a 13 A fuse is required.

Improve your grade

Electrical safety

Explain how the fuse and earth wire operate to protect a user. *AO1* [2 marks]

Ultrasound

Longitudinal waves

- **Ultrasound** is sound above 20 000 **Hz** which is a higher **frequency** than humans can hear.
  - It travels as a **pressure wave** containing **compressions** and **rarefactions**.
- Compressions are regions of higher pressure and rarefactions are regions of lower pressure.

Remember!
All sound, including ultrasound, is produced by vibrating particles.

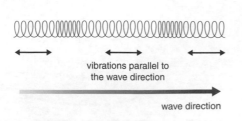

vibrations parallel to the wave direction

wave direction

A longitudinal wave

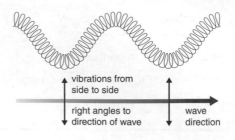

vibrations from side to side

right angles to direction of wave

wave direction

A transverse wave

- The features of **longitudinal** sound waves are:
  - They can't travel through a **vacuum**. The denser the medium, the faster a sound **wave** travels.
  - The higher the frequency or **pitch**, the smaller the **wavelength**.
  - The louder the sound, or the more powerful the ultrasound, the more energy is carried by the wave and the larger its **amplitude**.
- In a longitudinal wave the vibrations of the particles are parallel to the direction as the wave.
- In a **transverse** wave the vibrations of the particles are at right angles to the direction of the wave.

Uses of ultrasound

- When ultrasound is used to break down kidney stones:
  - a high-powered ultrasound beam is directed at the kidney stones
  - the ultrasound energy breaks the stones down into smaller pieces
  - the tiny pieces are then excreted from the body in the normal way.
- When ultrasound is used in a body scan, a pulse of ultrasound is sent into the body.
  - At each boundary between different tissues some ultrasound is **reflected** and the rest is transmitted.
  - The returning **echoes** are recorded and used to build up an image of the internal structure.
- Ultrasound can be used for body scans because:
  - when ultrasound is reflected from different interfaces in the body, the depth of each structure is calculated by using the formula distance = **speed** × time, knowing the speed of ultrasound for different tissue types and the time for the echo to return
  - the proportion of ultrasound reflected at each interface depends on the densities of each of the adjoining tissues and the speed of sound in the adjoining tissues
  - if the tissues are very different (e.g. blood and bone) most of the ultrasound is reflected, leaving very little to penetrate further into the body
  - the information gained is used to produce an image of the part of the body scanned.
- Ultrasound is preferred to **x-rays** because:
  - it is able to produce images of soft tissue
  - it doesn't damage living cells.

Improve your grade

Ultrasound and imaging

(a) State one similarity and one difference between sound and ultrasound waves.

(b) Give two reasons why doctors may decide to use ultrasound instead of x-rays to get images of inside the body. *AO1* [4 marks]

What is radioactivity?

Radioactive decay

- **Radioactive** substances decay naturally, giving out **alpha**, **beta** and **gamma** radiation.
- Nuclear radiation causes **ionisation** by removing **electrons** from **atoms** or causing them to gain electrons.
- Radioactive decay is a **random** process; it isn't possible to predict exactly when a nucleus will decay.
- There are so many atoms in even the smallest amount of **radioisotope** that the average **count rate** will always be about the same. Radioisotopes have unstable nuclei. Their nuclear particles aren't held together strongly enough.
- The **half-life** of a radioisotope is the average time for half the nuclei present to decay. The half-life cannot be changed.

The nucleus

- A **nucleon** is a particle found in the **nucleus**. So, **protons** and **neutrons** are nucleons.
- The nucleus of an atom can be represented as:

  $^A_Z X$ where
  - A = atomic mass (or nucleon number)
  - Z = **atomic number** (or proton number)
  - X = chemical symbol for the element

  Z = the number of protons in the nucleus, so the number of neutrons = (A – Z).

  The carbon **isotope** $^{14}_6 C$ has 6 protons and 8 neutrons in its nucleus.

Remember!
Nucleons cannot be lost. Charge is always conserved.

What are alpha and beta particles?

- When an **alpha** or a **beta particle** is emitted from the nucleus of an atom, the remaining nucleus is a different element.

- Alpha particles are very good ionisers. They are the largest particles emitted in radioactive decay. This means they are more likely to strike atoms of the material they are passing through, **ionising** them.

	alpha (α) particle	beta (β) particle
properties	positively charged has a large mass is a **helium** nucleus has helium gas around it consists of 2 protons and 2 neutrons	negatively charged has a very small mass travels very fast is an electron
during decay	**mass number** decreases by 4 nucleus has two fewer neutrons nucleus has two fewer protons atomic number decreases by 2	mass number is unchanged nucleus has one less neutron nucleus has one more proton atomic number increases by one
nuclear equation for decay	$^{238}_{92}U \rightarrow ^{234}_{90}Th + ^4_2 He$	$^{14}_6 C \rightarrow ^{14}_7 N + ^0_{-1}e$

EXAM TIP
You must know the difference between alpha, beta and gamma radiation.

Improve your grade

Nuclear radiation
When carbon-14 undergoes beta decay what happens to the atom and what new element is formed? *AO1* [2 marks]

Uses of radioisotopes

Background radiation

D–C

- Background **radiation** is due to:
  - **radioactive** substances present in rocks (especially **granite**) and soil
  - **cosmic rays** from Space
  - man-made sources including **radioactive waste** from industry and hospitals.

B–A*

- Most **background radiation** is from natural sources but some comes from human activity. This is shown in the pie chart.

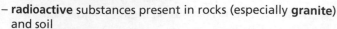

Remember!
Always subtract background radiation from any measurement of activity of a radioactive source.

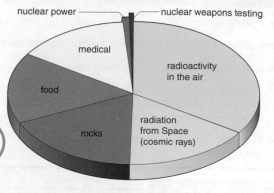

Sources of background radiation

Tracers

D–C

- When using a **tracer** to locate a leak in an underground pipe:
  - a very small amount of a **gamma** emitter is put into the pipe
  - a detector is passed along the ground above the path of the pipe
  - an increase in activity is detected in the region of the leak and little or no activity is detected after this point.

B–A*

- Gamma sources are used as tracers because the radiation is able to penetrate to the surface.

Smoke detectors

D–C

- A **smoke detector** contains an **isotope** which emits **alpha particles**.
  - Without smoke, the alpha particles **ionise** the air which creates a tiny **current** that can be detected by the circuit in the smoke alarm.
  - With smoke, the alpha particles are partially blocked so there is less ionisation of the air. The resulting change in current is detected and the alarm sounds.

Dating rocks

D–C

- Some rock types such as granite contain traces of **uranium**, a radioactive material.
  - The uranium isotopes present in the rocks go through a series of decays, eventually forming a **stable** isotope of **lead**.
  - By comparing the amounts of uranium and lead present in a rock sample, its approximate age can be found.

B–A*

- Uranium-238 decays, with a very long **half-life** of 4500 million years.
  - $^{238}_{92}U \rightarrow {}^{234}_{90}Th \ (+ {}^{4}_{2}He) \rightarrow {}^{234}_{91}Pa \ (+ {}^{0}_{-1}e) \rightarrow \ldots\ldots\ldots \rightarrow {}^{206}_{82}Pb$ (stable)
- The proportion of lead increases as time increases. If there are equal quantities of $^{238}_{92}U$ and $^{206}_{82}Pb$, the rock is 4500 million years (one half-life) old.

Radiocarbon dating

D–C

- **Carbon-14** is a radioactive isotope of carbon that is present in all living things. By measuring the amount of carbon-14 present in an archaeological find, its approximate age can be found.

B–A*

- Carbon dating can only be used on objects that were once alive.
  - When an object dies, no more carbon-14 is produced.
  - As the carbon-14 decays, the activity of the sample decreases.
  - The ratio of current activity from living matter to the sample activity provides a reasonably accurate date.

Improve your grade

Radioisotope dating
Carbon-14 has a half-life of 5700 years.

(a) What is meant by the half-life of a radioactive sample?

(b) A sample of bone was found to have a 25% of the amount of carbon-14 found in a living organism. How old was this bone?

AO1 [1 mark], *AO2* [2 marks]

Treatment

Using radiation

- Radiation emitted from the **nucleus** of an unstable **atom** can be **alpha** (α), **beta** (β) or **gamma** (γ).
  - Alpha radiation is absorbed by the skin so is of no use for diagnosis or **therapy**.
  - Beta radiation passes through skin, but not bone. Its medical applications are limited but it is used, for example, to treat the eyes.
  - Gamma radiation is very penetrating and is used in medicine. Cobalt-60 is a gamma-emitting **radioactive** material that is widely used to treat **cancers**.
- When nuclear radiation passes through a material it causes **ionisation**. Ionising radiation damages living cells, increasing the risk of cancer.
- Cancer cells within the body can be destroyed by exposing the affected area to large amounts of radiation. This is called **radiotherapy**.
- Materials can be made radioactive when their **nuclei** absorb extra **neutrons** in a nuclear reactor.

Comparing x-rays and gamma rays

- When **x-rays** pass through the body the tissues absorb some of this ionising radiation. The amount absorbed depends on the thickness and the **density** of the absorbing material.
- Gamma rays and x-rays have similar **wavelengths** but are produced in different ways.
- X-rays are made by firing high-**speed electrons** at metal targets.
- An x-ray machine allows the rate of production and **energy** of the x-rays to be controlled, but you can't change the gamma radiation emitted from a particular radioactive source.
- When the nucleus of an atom of a radioactive substance decays, it emits an alpha or a beta particle and loses any surplus energy by emitting gamma rays.

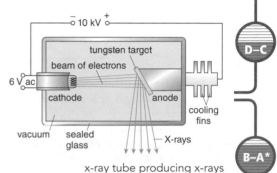

x-ray tube producing x-rays

Tracers

- A radioactive **tracer** is used to investigate inside a patient's body without surgery.
  - Technetium-99m is a commonly used tracer. It emits only gamma radiation.
  - **Iodine**-123 emits gamma radiation. It is used as a tracer to investigate the **thyroid gland**.
  - The radioactive tracer being used is mixed with food or drink or injected into the body.
  - Its progress through the body is monitored using a detector such as a gamma camera connected to a computer.

Treating cancer

- A **radioisotope** is used to destroy a **tumour** in the body.
- Three sources of radiation, each providing one-third of the required dose, are arranged around the patient with the tumour at the centre.
  - The healthy tissue only receives one-third of the dose, which limits damage to it.
- Or each radiation source is slowly rotated around the patient. The tumour receives constant radiation but healthy tissue receives only intermittent doses.

Treating a brain tumour with gamma radiation

Improve your grade

Medical tracers

Explain what a medical tracer is and how suitable materials are chosen to be used as tracers. *AO1* [3 marks]

Fission and fusion

Nuclear power stations

- Natural **uranium** consists of two isotopes, uranium-235 and uranium-238.
  - The 'enriched uranium' used as fuel in a **nuclear power station** contains a greater proportion of the uranium-235 **isotope** than occurs naturally.
- **Fission** occurs when a large **unstable nucleus** is split up and **energy** is released as heat.
  - The heat is used to boil water to produce steam.
  - The pressure of the steam acting on the **turbine** blades makes it turn.
  - The rotating turbine turns the **generator**, producing electricity.
- When uranium fissions, a **chain reaction** starts. A nuclear bomb is an example of a chain reaction that is not controlled.

- In a nuclear power station, **atoms** of uranium-235 are bombarded with **neutrons**. This causes the nucleus to split, releasing energy.
- A typical fission can be shown as:

$$^{235}_{92}U + {}^{1}_{0}n \rightarrow {}^{90}_{36}Kr + {}^{143}_{56}Ba + 3({}^{1}_{0}n) + \gamma\text{-rays}$$

- The extra neutrons emitted cause further uranium nuclei to split. This is described as a chain reaction and produces a large amount of energy.

a neutron is absorbed by the nucleus of a uranium-235 atom

the nucleus is now less stable than before

it splits into two parts and releases energy

several neutrons are also produced – these may go on to strike the nuclei of other atoms causing further fission reactions

this is called a chain reaction

Uranium-235 undergoing a chain reaction

Controlling nuclear fission

- The output of a nuclear reactor can be controlled.
  - A **graphite moderator** between the **fuel rods** slows down the fast-moving neutrons emitted during fission. Slow-moving neutrons are more likely to be captured by other uranium nuclei.
  - **Boron control rods** can be raised or lowered. Boron absorbs neutrons, so fewer neutrons are available to split more uranium nuclei. This controls the rate of fission.

Boron control rods can be lowered into the reactor to absorb neutrons

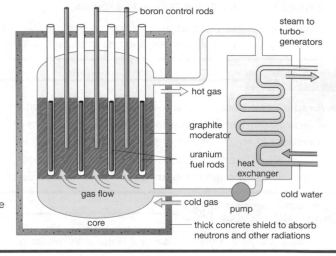

boron control rods

steam to turbo-generators

hot gas

graphite moderator

uranium fuel rods

heat exchanger

cold water

gas flow

cold gas

pump

core

thick concrete shield to absorb neutrons and other radiations

Fusion

Nuclear fusion happens when two light nuclei fuse (join) together releasing large amounts of heat energy.

- Fusion requires extremely high temperatures which have proved difficult to achieve and to manage safely on Earth and so its use for large-scale power generation remains a dream.
- Research in this area is very expensive so it is carried out as an international joint venture to share costs, expertise and benefits.

- In stars, fusion takes place under extremely high temperatures and pressures.
- **Fusion bombs** are started with a fission reaction which creates exceptionally high temperatures.
- So far attempts to replicate these conditions safely on Earth have been unsuccessful.
  - Scientists are still trying to solve the safety and practical challenges presented.
  - **Cold fusion** is still not accepted as realistic since any results are impossible to verify so far.

Improve your grade

Controlling nuclear fission
Explain how the output of a nuclear reactor is controlled. *AO1* [3 marks]

P4 Summary

Electrostatics

The chance of getting an electric shock can be reduced by:

- correct earthing
- standing on insulating mats
- wearing shoes with insulating soles
- bonding fuel tanker to aircraft.

Electrostatic effects are caused by the transfer of electrons.
A positively charged object lacks electrons.
A negatively charged object has extra electrons.

There are two kinds of electric charge, positive and negative:

- like charges repel
- unlike charges attract.

Uses of electrostatics include:
Defibrillators, paint and crop sprayers, dust precipitators and photocopiers.

Using electricity safely

In a three pin plug:

- live wire is at a high voltage
- neutral wire completes the circuit
- earth wire is connected to the case to prevent it becoming live.

Double insulated appliances do not need an earth wire.

power = voltage × current
The fuse (or circuit breaker) is in the live wire. If the current is greater than the rating, the fuse will melt (or circuit breaker switch) breaking the circuit.

$$resistance = \frac{voltage}{current}$$

Longer wires have more resistance.
Thinner wires have more resistance.

Ultrasound

Medical uses include:

- scans to see inside the body without surgery
- measuring the rate of blood flow
- breaking up kidney stones.

Ultrasound is sound above 20 000 Hz which is a higher frequency than humans can hear.

Longitudinal waves – e.g. sound and ultrasound.
Transverse waves – e.g. light.
Wavelength (λ) is the distance occupied by one complete wave.
Frequency (f), measured in hertz (Hz), is the number of complete waves in 1 second.

Nuclear radiation

Nuclear equations are used to represent α decay:

$$^{238}_{92}U \rightarrow {}^{234}_{90}Th + {}^{4}_{2}He$$

or β decay:

$$^{14}_{6}C \rightarrow {}^{14}_{7}N + {}^{0}_{-1}e$$

Gamma rays are given out from the nucleus of certain radioactive materials.
x-rays are made by firing high speed electrons at metal targets.

Nuclear radiation is emitted from the nuclei of radioisotopes:

- α particle is a helium nucleus
- β particle is a fast-moving electron
- γ radiation is a short wavelength electromagnetic wave.

Medical uses of radioactivity include:

- diagnosis, as a tracer
- sterilising equipment
- treating cancers.

Only β and γ can pass through skin. γ radiation is most widely used as it is the most penetrating.

Other uses include:

- smoke detectors
- industrial tracers
- dating rocks and archaeological finds.

Fission is the splitting up of a large nucleus.
Fusion is the joining together of smaller nuclei.
Both release a lot of energy.
Fission is currently used to generate electricity.
Fission leads to a chain reaction which must be carefully controlled.

The half-life is the average time for half of the nuclei present to decay.

Satellites, gravity and circular motion

Gravity

- Every object in the Universe attracts every other object.
- These forces of attraction are usually only significant when they are on an astronomical scale.
- Planets stay in **orbit** around the Sun because of **gravitational attraction**.
- Any object moving in a circle needs a force towards the centre of the circle. This force is called **centripetal force**.

Gravitational forces

- The larger the mass, the greater the **gravitational** force.
- The further away an object is, the smaller the gravitational force.
- Gravitational forces obey an inverse square law.

$$\text{force} \propto \frac{1}{\text{distance}^2}$$

- The orbit time for a planet closer to the Sun is less because:
  - the planet travels a shorter distance
  - the planet travels faster because there is a greater gravitational force.
- A **satellite** moves at a tangent and **gravity** makes it **accelerate** towards Earth.
- **Comets** travel very quickly when close to the Sun and very slowly when they are far away.

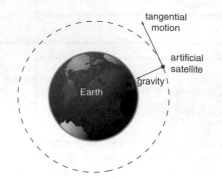

A satellite travels at a constant speed around the Earth but accelerates towards the Earth

Satellite orbits

- **Polar orbit** satellites orbit above the North and South poles just 100 km – 20 km above Earth.
- Most polar orbit satellites take about 90 minutes to orbit the Earth.
- Polar orbit satellites can see the whole of the Earth's surface as the Earth rotates beneath them.
- They are used for imaging the Earth – including short-**range** weather forecasting.
- A **geostationary satellite** orbits above the equator.
- The time for one orbit is 24 hours – this means it always appears to be in the same position above the equator.
- Geostationary satellites are used for communications.
- The nearer a satellite is to the Earth's surface, the shorter its orbit time.
- The distance travelled by a satellite is $2\pi r$ where r is the height of the satellite above the centre of the Earth.

$$\text{time} = \frac{\text{distance}}{\text{speed}} = \frac{2\pi r}{\text{speed}}$$

- If r is small and **speed** is large then time will be small – this is a polar orbit.
- All geostationary satellites must be in the same orbit.
  - They cannot be too close together as their signals would overlap due to diffraction.

Improve your grade

Satellite speed

Calculate the approximate speed of a satellite in low polar orbit. Assume the radius of the Earth is 6400 km, the altitude of the satellite 100 km and the time taken to orbit 90 minutes. *AO1* [1 mark] *AO2* [2 marks]

Vectors and equations of motion

Scalars and vectors

- A **scalar** quantity has **magnitude** only – a **speed** of 20 m/s is a scalar.
- A **vector** quantity has both magnitude and direction – a speed of 20 m/s due north is a vector.
- Speed in a particular direction is called **velocity**.
- Vectors are usually represented by arrows in the correct direction; the length of the arrow represents the magnitude of the vector.
  3 m/s to the right ⟶ 2 m/s to the left ⟵
- When vectors act in a straight line, they are added algebraically.
  3 m/s to the right ⟶ + 2 m/s to the left ⟵ = 1 m/s to the right ⟶

Adding non-linear vectors

- The resultant of two forces that are not in the same straight line is found using the **parallelogram of forces**.

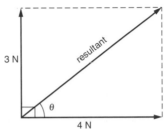

The resultant is 5 N acting at 36.9° to the 4 N force

- If the two forces are at right angles, the parallelogram is a rectangle.
- The two forces are represented in size and direction by the adjacent sides of a parallelogram.
- The resultant is represented in size and direction by the diagonal of the parallelogram from the same point.

Equations of motion

- We use symbols to represent five different quantities:
  u = initial velocity in m/s
  v = final velocity in m/s
  a = acceleration in m/s²
  s = distance travelled in m
  t = time taken in s
- Final velocity can be calculated using: $v = u + at$
- Distance travelled can be calculated using: $s = \frac{1}{2}(u + v)t$
- These equations can be combined to produce two other equations: $s = ut + \frac{1}{2}at^2$
                                                                     $v^2 = u^2 + 2as$
- If three of the quantities u, v, a, s and t are known, the other two can be found.
- All bodies fall to Earth under the effect of **gravity**.
- They **accelerate** at 10 m/s².
- An object thrown into the air slows down with an acceleration a = –10 m/s².
- At the top of its ascent, its velocity is 0.

EXAM TIP

When using equations of motion, make sure you always work in the correct units.

Improve your grade

Scalars and vectors

(a) Explain the difference between a scalar and a vector. *AO1* [1 mark]
(b) Which of the following are vectors and which are scalars?
    mass   temperature   force   acceleration   volume   *AO1* [2 marks]

Projectile motion

Horizontal projection

- A ball thrown horizontally from the top of a tower will fall to the ground at the same rate as a ball dropped from the top of the tower.

- The **trajectory** of the ball thrown horizontally is **parabolic**.

- There is no horizontal force on the ball (if you neglect air resistance).

- The ball has a constant horizontal **velocity**.

- Both balls **accelerate** towards the ground.

- Newton suggested that if the ball could be kicked hard enough from a very tall peak, it would never hit the ground but stay in **orbit**. **Gravity** would make the ball fall to Earth but never reach it because of the Earth's curvature.
  - This is the principle used to put **satellites** into orbit.

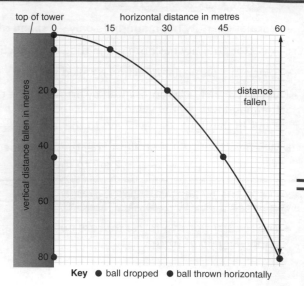

How two balls behave as they fall to the ground

Newton's drawing from Principia published 1687

Projectile calculations

- The four equations of motion can be applied to **projectiles**:

$v = u + at$

$s = \frac{1}{2}(u + v)t$

$s = ut + \frac{1}{2}at^2$

$v^2 = u^2 + 2as$

EXAM TIP

Apply the equations separately to the horizontal (constant velocity) motion and the vertical motion.

Analysing projectile motion

- The photograph taken with a stroboscopic light shows:
  - the horizontal **speed** is constant
  - the ball slows down as it goes up
  - the ball speeds up as it falls.

- The only force on the ball is a vertical force due to gravity.

- force = mass × acceleration so there is a downwards acceleration of 10 m/s².

A stroboscopic light takes a picture of a bouncing ball at regular intervals

Resultant velocity

- Velocity is a vector – it has **magnitude** and direction.

- The resultant velocity is the **vector** sum of the constant horizontal velocity, v_H and the vertical velocity, v_V.

- These vectors are at right angles so the resultant is represented in magnitude and direction by the diagonal of the vector rectangle.

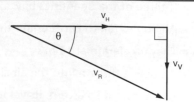

Use the parallelogram of vectors to calculate the resultant velocity

Improve your grade

Projectiles

Explain why a satellite in orbit can be thought of as being like a projectile. *AO2* [2 marks]

Action and reaction

Force pairs

- Forces always occur in pairs.
- The **action** and reaction pairs of forces:
  - are equal in **magnitude** (size)
  - are opposite in direction
  - act on different objects.
- This is Newton's third law.
- The boy has **weight** because he is attracted to Earth due to **gravity**.
- The Earth is also attracted to the boy; this is the **reaction force**.
- If the boy jumps in the air he is attracted to Earth and moves down to it.
- Earth is also attracted to the boy with an equal force but its mass is so great that it only moves a very small amount, much too small for us to notice.
- When a gun is fired, there is a force on the bullet from the gun so it moves forward and the gun **recoils** due to the force from the bullet.

The boy is attracted to the Earth and the Earth is attracted to the boy

Conservation of momentum

- The **momentum** of an object depends on its mass and **velocity**.

  momentum = mass × velocity
- In any collision, the total momentum before the collision is equal to the total momentum after the collision.

  $m_1 u_1 + m_2 u_2 = (m_1 + m_2) v$
- An explosion is the opposite of a collision. Momentum is still conserved.

Kinetic theory of gases

- Every material is made of tiny moving particles with **kinetic energy**.
- The higher the temperature the more kinetic energy the particles have and the faster they move.
- In a gas the particles are far apart and free to move.
- The particles collide with the walls of the container, creating a force, and hence a **pressure**, on the walls.
- Changing volume – if the same number of particles, at the same temperature, are in a smaller container they hit the walls more often, increasing the pressure.
- Changing temperature – if the same number of particles in the same container are heated, the particles move faster hitting the walls more often and with greater force, increasing the pressure.

Gas pressure

force = rate of change of momentum

- Consider a particle moving in a sealed box:
  - the particle has mass, m, moving with velocity, –v, and momentum –mv
- Each time the particle collides perpendicularly with a wall it rebounds with the same **speed** but in the opposite direction. Its velocity is +v and its momentum is +mv.
  - Change of momentum is mv – (– mv) = 2 mv.
  - Force on wall, F = 2 mv/t where t = time between collisions with wall.
- As a **rocket** moves up, the hot gases released move down. Momentum is conserved. This means that the high momentum of the large massed rocket moving up is balanced by the high velocity of the exhaust gases.

Improve your grade

Explosions

When a radioactive nucleus emits a beta particle, the nucleus moves slowly in the opposite direction to the fast moving beta particle. Explain why. *AO2* [2 marks]

Satellite communication

Communicating with satellites

D–C

- **Microwave** signals are sent into space from a **parabolic transmitter**.
- The signals are received, **amplified** and re-transmitted back to Earth by a **geostationary satellite**.
- The signals are picked up by a parabolic **receiver**.
- **Digital** signals are used for **satellite** communication.
  - They do not **attenuate** as quickly.
  - There is less **noise**.
- The **ionosphere** reflects **radio waves** with **frequencies** below 30 MHz.
- Frequencies above 30 GHz are **absorbed** and **scattered**. This reduces signal strength.
- The frequencies used for satellite communication are between 3 GHz and 30 GHz.

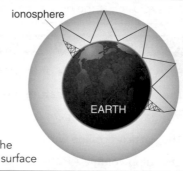

Remember!
mega means million (10^6), giga means thousand million (10^9).

B–A*

- Geostationary satellites **orbit** Earth about 36 000 km above the equator.
  - The size of the **aerial** dish is many times the microwave **wavelength** so there is very little **diffraction**. It produces a narrow beam that does not spread out.
  - This means the receiving dish and satellite dish need exact alignment to ensure that the signals do not 'miss' the geostationary satellite.
- All geostationary satellites are in the same orbit which is very crowded.
- **Microwaves** pass through the Earth's atmosphere and because they have a very small wavelength they do not spread out very much.
- The narrow microwave beams are able to target one satellite without interfering with another.

The ionosphere

B–A*

- The **ionosphere** is a region between 100 km and 500 km above the Earth's surface.
- Radio waves undergo a series of **refractions** and **speed** up as they enter different layers. Eventually, the waves are totally internally reflected.
- Radio waves also reflect off the Earth's surface. This means they can travel around the world.

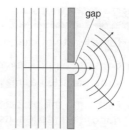

Radio waves are reflected from the ionosphere and from the Earth's surface

Diffraction

D–C

- The smaller the size of a gap, the greater the diffraction.
- Long and medium wave radio waves diffract around buildings, hills and follow the curvature of the Earth.
- TV signals have shorter wavelengths so do not show much diffraction. Television **aerials** must be in line of sight with the **transmitter**.

B–A*

- Maximum diffraction occurs when the wavelength is equal to the size of the gap.
- Radio waves have long **wavelengths** compared to the distance between hills so diffract easily.

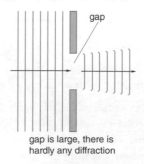

gap is large, there is hardly any diffraction

gap is small, waves diffract

Diffraction of waves as they pass through a gap

Improve your grade

Geostationary satellites

Geostationary satellites are used for communication. Explain why the transmitting and receiving dishes need very careful alignment and why the satellites must be as far away from each other as possible. *AO2* [3 marks]

Nature of waves

Interference

- If two loudspeakers are connected to the same output of an oscillator and placed about 1 m apart, you will hear alternate loud and quiet areas as you move along a line in front of them.

- When two **crests** or two **troughs** overlap, the waves are in step and the sound will be loud.

- When a trough and a crest overlap, the waves are out of step and the sound will be quiet.

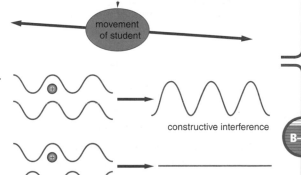
oscillator

loudspeakers

1 m

movement of student

Interference can be heard if you walk across in front of two loudspeakers producing the same **frequency** note

D–C

- **Constructive interference** means that the waves meet **in phase**.

- **Destructive interference** means that the waves are out of phase.

- Constructive **interference** occurs when the path difference between the two sources is a whole number of **wavelengths**.

- Destructive interference occurs when the path difference between the two sources is an odd number of half wavelengths.

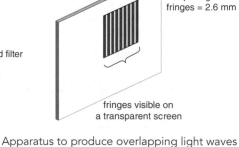
constructive interference

destructive interference

Constructive and destructive interference

B–A*

Light as a wave

- Interference between light waves is produced when light diffracts as it passes through narrow slits.
  - Bright bands are constructive interference.
  - Dark bands are destructive interference.

single narrow slit

light bulb

two narrow slits (about 0.1 mm wide) separated by about 0.25 mm

red filter

1 m

spacing of fringes = 2.6 mm

fringes visible on a transparent screen

Apparatus to produce overlapping light waves

D–C

- The wavelength of light is between 0.0004 mm and 0.0006 mm so the slits have to be very narrow.

B–A*

- Interference of light can only be explained if light is a wave.

- **Electromagnetic waves**, such as light, are **transverse** waves. Oscillations occur in all directions at right angles to the wave direction.

- Light is **polarised** if the oscillations are only in one direction at right angles to the wave direction.

D–C

- **Polaroid** sunglasses stop oscillations in all but one direction at right angles to the wave. This reduces the amount of light passing through.

- Newton thought of light as a particle. If he was correct, light should travel faster in a denser medium.

- Huygens thought of light as a wave. If he was correct, light should travel slower in a denser medium.

- When the **speed** of light was finally measured, Newton was proved wrong.

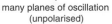
many planes of oscillation (unpolarised)

one plane of oscillation (polarised)

Polarisation

B–A*

Improve your grade

Interference

An oscillator, connected to two loudspeakers, is producing a sound with a wavelength of 0.4 m. Jane stands 5.5 m away from one loudspeaker and 7.7 m away from the other. Explain what she will hear. *AO2* [4 marks]

Refraction of waves

Explaining refraction

D–C

- **Refraction** occurs because when light enters a different medium, its **speed** changes:
  - light slows down as it enters a more dense medium – it deviates towards the normal
  - light speeds up as it enters a less dense medium – it deviates away from the normal.
- The **refractive index** indicates the amount of deviation – the greater the deviation, the higher the refractive index.

$$\text{refractive index (n)} = \frac{\text{speed of light in vacuum}}{\text{speed of light in medium}}$$

- **Dispersion** occurs because each colour slows down by a different amount when white light enters a medium such as glass, plastic or water, and speeds up by a different amount on leaving.

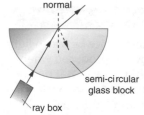

Refraction through a glass block

B–A*

- Light waves are refracted when they enter an optically denser medium because they slow down – their **wavelength** is smaller.

Snell's Law states that:

$$\text{refractive index} = \frac{\text{sin angle of incidence}}{\text{sin angle of refraction}}$$

- Dispersion occurs because different colours have different speeds in a medium and have different refractive indices. This means each colour has a different **angle of refraction**.

Critical angle

D–C

- When light passes from a more dense medium into a less dense medium, the angle of refraction is larger than the **angle of incidence**.
- The **critical angle** is the angle of incidence in the more dense medium that produces an angle of refraction of 90° in the less dense medium.
- If the critical angle is exceeded, the light is totally internally **reflected**.
- **Optical fibres** rely on **total internal reflection** transmitting light along a thin fibre.
- Optical fibres are used:
  - to carry telephone conversations and computer data as pulses of laser light
  - in endoscopes to look inside the body without surgery.

B–A*

- Total internal reflection only happens when:
  - a ray of light travels from one medium towards another with a lower refractive index (for example, glass to air or glass to water) so that the angle of refraction reaches 90° as the angle of incidence is increased
  - the angle of incidence is greater than the critical angle.
- The higher the refractive index of a material, the lower its critical angle.

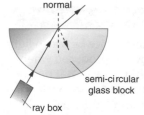

When the angle of incidence is less than the critical angle, c, most of the light is refracted out of the glass block.

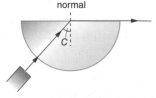

i = c and the angle of refraction = 90°

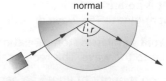

When the angle of incidence is greater than the critical angle, all the light is totally internally reflected.

Behaviour of light below, at and greater than the critical angle

Improve your grade

Dispersion

Explain why white light passing into a prism leaves a spectrum of colours.
AO1 [2 marks]

Optics

Focal points

- A parallel beam of light from a distant object can be **converged** to a **focus** in the **focal plane**.
- If the beam is parallel to the **principal axis**, the light is focused to the **focal point** on the principal axis.
- A **diverging** beam from a near object will focus at a point beyond the **focal plane**.
- A **camera** is a box with a **lens** at one end and a film at the other.
  - Light from an object passes through the lens and is focused on the film.
  - The size of the image is smaller than the object.
- In a **projector**, the film is placed closer to the lens than the object is to the camera.
  - Light from a bulb passes through the film and the lens and is focused on the screen.
  - The size of the image is larger than the object.

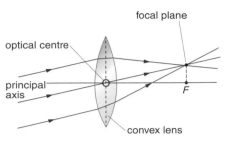

The action of a convex (**converging**) lens

D–C

Ray diagrams

- The position and size of an image formed by a **convex** lens can be found using scale diagrams.
- The paths of two rays from the top of the object are as follows:
  - the ray parallel to the principal axis **refracts** through the focal point
  - the ray through the centre of the lens is not deviated.
- Where the rays meet is at the top of the image.

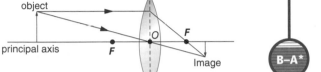

object more than twice the focal length from lens

Using a scale diagram to find the size and position of an image

B–A*

Using convex lenses

- The image formed by a magnifying glass is the right way up and cannot be projected onto a screen.
- A camera forms a real, inverted, small image on the film.
- Simple lens cameras have a fixed lens. This only produces a sharp image for one object distance. Better quality cameras have an adjustable lens distance. This means the image is sharp whatever the object distance.
- The **shutter** opens and closes to allow light on to the film.
- The **aperture** is an adjustable hole allowing different amounts of light onto the film.
- Projectors form a large, inverted, **real image** on a screen.
  - The lens and/or screen can be moved to produce a sharp image.
  - The **condenser lenses** make sure the film is uniformly illuminated.
  - The curved mirror reflects light back to the condenser lenses.

D–C

Magnification

$$\text{magnification} = \frac{\text{image size}}{\text{object size}}$$

D–C

Virtual images

- The image formed by a magnifying glass is a **virtual image** – it cannot be projected onto a screen because no light passes through it.

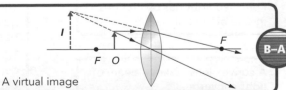

A virtual image

B–A*

Improve your grade

Camera lenses

A simple camera has a fixed lens.

(a) Draw a ray diagram to show where the image is formed if the object is a long way away from the camera.

(b) How far from the lens is the film in the camera? *AO2* [3 marks]

Satellites and gravity

Geostationary satellites take exactly 24 hours to orbit Earth, above the equator. This means they always stay above the same point on Earth.

A gravitational force keeps a satellite in orbit.

The height and period of a satellite's orbit depends on its job.

Any object moving in a circle needs a force towards the centre of the circle to maintain its circular path.

This is called a centripetal force.

Satellite communication

Some radio waves are reflected by the ionosphere.

High frequency radio waves (microwaves) pass through it to reach an orbiting satellite.

Long wavelength radio waves are easily diffracted around hills; short wavelength microwaves only diffract a small amount.

The amount of diffraction depends on the size of the gap or obstacle.

Scalars and vectors

A scalar quantity (e.g. speed) has size only.

A vector quantity (e.g. velocity) has size and direction.

Vector diagrams are used to calculate the resultant forces.

Equations of motion (for uniform acceleration):

$$v = u + at$$
$$s = \frac{(u + v)\,t}{2}$$
$$v^2 = u^2 + 2as$$
$$s = ut + \tfrac{1}{2}at^2$$

Changes in temperature and volume cause changes in pressure.

Rate of change of **momentum** creates a force on the walls of the container.

Rockets expel a large number of particles at high speed.

Projectiles:
- have a constant horizontal velocity
- have a trajectory that is parabolic
- accelerate towards the ground at $10\ \text{m/s}^2$.

momentum = mass × velocity

Momentum is a vector.

Momentum is always conserved.

Waves

Light travels in straight lines.

Diffraction and interference of light can only be explained by a wave model.

Polarised light has oscillations in one plane only.

Interference occurs when two waves overlap. They can reinforce or cancel.

This results in louder and quieter areas in sound and bright and dark areas in light.

The type of interference depends on the path difference between the two waves.

Total internal reflection can occur when light goes from a dense to a less dense medium. The critical angle is the angle of incidence for which the angle of refraction is 90°. At greater angles of incidence all the light is totally internally reflected. This happens in an optical fibre.

A convex lens makes a beam of light converge.

Convex lenses are used in cameras and projectors to produce real, inverted images, and as a magnifying glass to produce a virtual image that is the right way up.

$$\text{magnification} = \frac{\text{image size}}{\text{object size}}$$

Refraction occurs when light passes from one medium to another. As the speed of the light changes, its direction can change.

The refractive index tells us about the amount of bending.

$$\text{refractive index} = \frac{\text{speed of light in vacuum}}{\text{speed of light in medium}}$$

Light sometimes produces a rainbow, or spectrum of colours – dispersion.

Resisting

Resistance and current

- The higher the **resistance**, the lower the **current**.
- The higher the current, the brighter the bulb in a circuit.
- The higher the current, the faster a motor turns.

  resistance = **voltage** ÷ current   current = voltage ÷ resistance   voltage = current × resistance

Factors affecting resistance

- The resistance of a length of wire depends on its length.
- A **variable resistor** works by changing the length of wire in the circuit; the longer the wire, the greater the resistance.

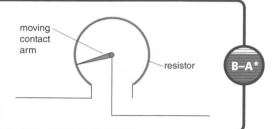

A dimmer switch has a contact that moves around a circular coil of wire

Ohmic and non-ohmic devices

Fixed resistor

- The voltage–current graph for a resistor is a straight line passing through the origin. This shows that voltage is directly proportional to current. A resistor obeys Ohm's law.
- The steeper the gradient of the voltage–current graph the greater the resistance.
- The voltage–current graph for a bulb is not a straight line passing through the origin. The gradient of the line increases as the current increases. A bulb does not obey Ohm's law.

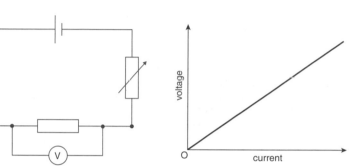

Voltage–current graph for a resistor

Filament lamp

- The resistance of a filament bulb increases as the current increases.
- When **electrons** collide with the atoms in the filament, it makes the atoms vibrate more. This increased vibration leads to:
  - an increased number of collisions so the resistance increases
  - an increase in the temperature of the filament.

- For a fixed resistor, the gradient of the graph is equal to the resistance.
- The gradient of the voltage–current graph for a filament lamp increases as the resistance increases. So the resistance must be found using instantaneous values from the graph.

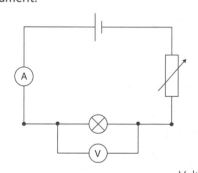

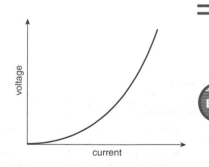

Voltage-current graph for a fixed resistor

Improve your grade

Resistance

Susan is using a 1 m long piece of nichrome wire to investigate the effect of length on resistance in the science laboratory. The teacher has chosen a piece of wire with a resistance of 100 Ω.

(a) Calculate the current passing when Susan puts 2 V across the ends of the wire.

(b) Sketch the graph you would expect Susan to have at the end of her investigation. *AO2* [3 marks]

Sharing

Potential divider circuits

D–C

- Some **potential divider** circuits have fixed **resistors**.
- The output **voltage** is a fixed proportion of the supply voltage.
- Some potential dividers have one **variable resistor**.
- The output voltage can be altered.

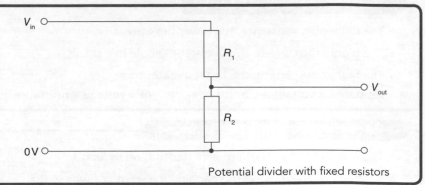

Potential divider with fixed resistors

Calculating output voltage

B–A*

$$V_{out} = \frac{R_2}{R_1 + R_2} \; V_2$$

- When R_2 is much larger than R_1, $V_{out} \approx V_{in}$.
- When R_2 is much less than than R_1, $V_{out} \approx 0$.
- Some electronic components will start working at a **threshold voltage**.
- Using a variable resistor as one part of the potential divider allows the threshold voltage to be set.

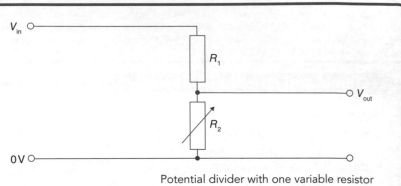

Potential divider with one variable resistor

Resistors in parallel

D–C

B–A*

- **Resistors** can be arranged in parallel and this will decrease the total resistance.
- Total resistance is found using

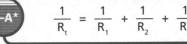

$$\frac{1}{R_t} = \frac{1}{R_1} + \frac{1}{R_2} + \frac{1}{R_3}$$

LDR and thermistor characteristics

D–C

- The resistance of a **light-dependent resistor** (LDR) decreases as the light intensity increases.
- The resistance of a **thermistor** decreases as the temperature increases.

Remember!

Remember the behaviour of a thermistor is opposite to that of a resistor – the resistance of a resistor increases as temperature increases.

resistance in Ω

light level in lux

LDR characteristics

Using potential divider circuits

B–A*

- Street lights have LDRs in potential divider circuits.
- When it is dark, the resistance of the LDR is high, the output voltage is high and switches on the streetlight.
- A thermistor can be used to switch on a heater when the temperature gets too low.

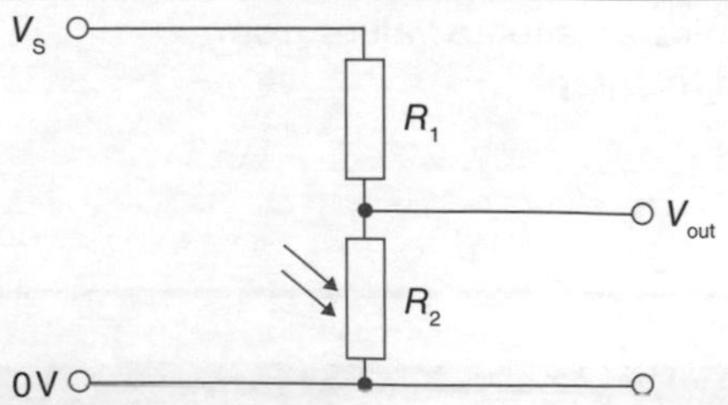

Circuit used in street lights

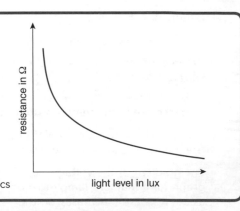

Improve your grade

Resistors in parallel

Calculate the total resistance when three 100 Ω resistors are connected in parallel.

AO2 [2 marks]

P6 Electricity for gadgets

It's logical

Transistors

- In a **transistor**, a small base **current** can switch on a greater current through the collector.

$$I_e = I_b + I_c$$

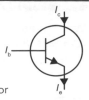

The symbol for a transistor

Making an AND gate

- Two transistors can be combined to make an **AND gate**.
  - A small input current at A will switch on the upper transistor.
  - A small input current at B will switch on the lower transistor.
  - A current at A *and* at B are needed to switch on the 6V supply.
- Other logic **gates** can be made from different combinations of transistors.
- Transistor circuits always have a high value **resistor** in the base circuit to limit the current.

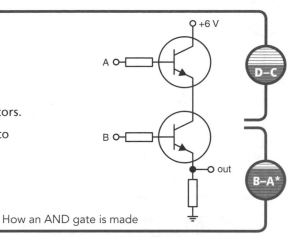

How an AND gate is made

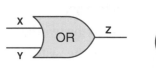

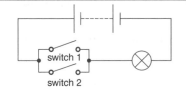

OR gate

- An **OR gate** behaves like two switches in parallel. The output is high if input X *or* input Y is high.

switch 1
switch 2

The symbol for an OR gate

NAND and NOR gates

- A **NAND gate** behaves as if it is an AND gate followed by a **NOT gate**.
- A **NOR** gate behaves as if it is an **OR** gate followed by a NOT gate.

The symbol for a NAND gate   The symbol for a NOR gate

Truth tables

- Truth tables show how logic gates behave.
  - When there is an input (or output), a 1 is entered in the table.
  - When there is no input (or output) a 0 is entered in the table.

- AND gate

Input A	Input B	out
0	0	0
0	1	0
1	0	0
1	1	1

OR gate

Input A	Input B	out
0	0	0
0	1	1
1	0	1
1	1	1

- NAND gate

Input A	Input B	out
0	0	1
0	1	1
1	0	1
1	1	0

NOR gate

Input A	Input B	out
0	0	1
0	1	0
1	0	0
1	1	0

Improve your grade

Logic gates
Draw a circuit diagram to show how an AND gate can be constructed from two transistors. *AO1* [3 marks]

Even more logical

Truth tables

- When there are several **logic gates** combined together, **truth tables** can be used to work out what happens in the system.

Input signals				Output
A	B	C	D	E
0	0	0	0	0
0	0	1	0	0
0	1	0	1	0
0	1	1	1	1
1	0	0	1	0
1	0	1	1	1
1	1	0	1	0
1	1	1	1	1

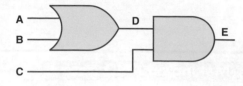

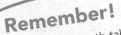

Logic circuit

- It does not matter how many logic gates there are when a truth table is being worked out.

Remember!
When you work out a truth table for a system, do it one step at a time. Write out all of the possible input values and then work out what happens at each logic gate.

Switching logic gates

- Switches, **LDRs** and **thermistors** are used in a **potential divider** circuit to make the input to a logic gate high.
- When the switch is open, the input to the logic gate is low.
- When the switch is closed, the input is connected directly to the 5 V supply. The input is high.
- The LDR and thermistor work in the same way.
- If a **variable resistor** is used instead of a fixed resistor, a threshold level can be set.

How a relay works

- When a **current** passes through the coil, the iron armature is attracted.
- The armature pivots and pushes an insulating bar against the central contact.
- The central contact moves, opening the normally closed contacts and closing the normally open contacts.

- Current output from logic gates is low but it can be passed through a **relay** to switch on a larger current needed for motors, heaters etc.
- The relay also isolates the logic circuit from the high current to avoid damage to the logic gates.

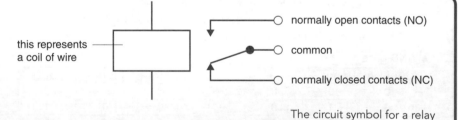

this represents a coil of wire

normally open contacts (NO)

common

normally closed contacts (NC)

The circuit symbol for a relay

Improve your grade

Truth tables

Tom builds a logic circuit with three logic gates as shown right. Draw a truth table for Tom's circuit. *AO2* [2 marks]

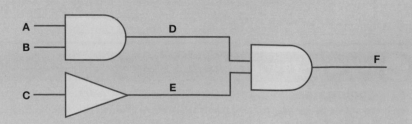

Motoring

Magnetic fields

- The direction of the field around a wire can be found by using the right-hand grip rule.
- The field pattern due to a long coil of wire is similar to that of a bar magnet.
- When a wire is placed between the poles of a magnet, the wire moves out of the gap when the **current** is switched on.
- If the direction of the current is reversed, the wire moves in the opposite direction.

Imagine gripping the wire with your right hand, with your thumb pointing in the direction of the current. Your fingers point around the wire in the direction of the field

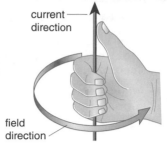

current direction

field direction

The right-hand grip rule

D–C

The polarity can be predicted from the current direction

north pole

end view

south pole

anticlockwise

clockwise

Remember!
You can work out the North and South ends of the coil by seeing which way the current passes.

Motor direction

- **Fleming's left-hand rule** allows you to predict the direction in which a motor turns.
- The direction of the current, **magnetic field** and motion are all at right angles to each other.

thumb - motion

first finger – field

second finger – current

Fleming's left-hand rule

B–A*

Turning coils

- When a current passes through a coil, placed between the poles of a magnet, there is a force on each side of the coil.
- Using Fleming's left-hand rule, you can see that the forces on each side of the coil are in opposite directions. As one side is forced up, the other side is forced down.
- The coil starts to spin.
- The motor spins faster when:
  - the number of turns on the coil is increased
  - the size of the current is increased
  - the strength of the magnetic field is increased.

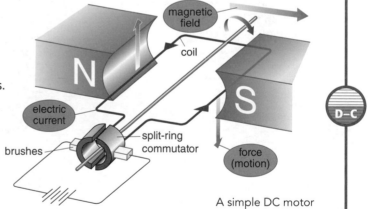

magnetic field

coil

N

S

electric current

brushes

split-ring commutator

force (motion)

A simple DC motor

D–C

Practical motors

- The job of the **commutator** is to make the motor continue to spin.
- The direction of the current in the coil is reversed every half turn.
- This makes sure the force on the coil is always in the same direction.
- The magnets in practical motors have curved poles to give a radial magnetic field.
- A radial field increases the force on the coil and keeps it constant as the motor turns.

N   S

coil

A radial field

B–A*

Improve your grade

DC electric motor

Jasmine has made a model electric motor. What three things can she change to make her motor spin faster? *AO1* [3 marks]

Generating

D–C

Induced voltages

- When a wire is held stationary between the poles of a magnet, there is no **current** in the wire.
- If the wire is moved upwards, a **voltage** is induced in the wire and a current passes. The same thing happens if the magnet is moved downwards. It is the relative movement of wire and field that is important.
- When the wire moves downwards or the magnet moves up, the induced voltage is reversed and the current is in the opposite direction.
- Reversing the direction of the **magnetic field** also reverses the induced voltage and current direction.
- A voltage is induced whenever a magnetic field changes.

B–A*

- The size of the induced voltage depends on the rate at which the magnetic field changes:
  - if the wire is moved slowly voltage is low
  - if the wire is moved quickly voltage is higher.

AC generators

D–C

- In a power station **generator**, the magnetic field inside the coil is produced by **electromagnets**. The electromagnet is made from coils of wire (rotor coils) which are turned by the turbine.
- An **alternating current** is produced in the stator coils which surround the rotor coils.
  - Increasing the **speed** of rotation of the electromagnet increases both the current induced in the stator coils and the **frequency** of the voltage generated.
  - Increasing the number of turns on the electromagnet increases the magnetic field and therefore a larger voltage is induced in the stator coils.

- Some AC generators do have a rotating coil between the poles of a magnet.
- As the coil rotates, the direction of the current in the coil reverses every half turn.
- **Slip rings** are connected to the ends of the coil to allow the coil to spin without winding the wire around itself.

B–A*

- The brushes are contacts that touch the slip rings and complete the circuit.

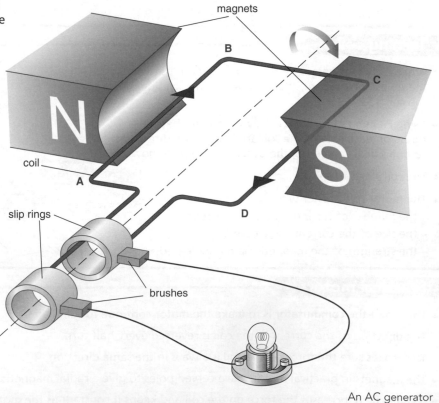

An AC generator

Improve your grade

Generators

Explain how an AC generator works including the action of the slip rings and brushes. *AO1* [4 marks]

Transforming

Transformer design

- A **transformer** consists of two coils of wire wound onto an iron core.
- The input AC **voltage** is connected to the **primary coil**.
- The output AC voltage is obtained from the **secondary coil**.
- Step-down transformers have more turns on the primary coil.
- Step-up transformers have more turns on the secondary coil.
- **Isolating transformers** are used for safety – the water and steam in the bathroom could lead to electrocution or damage to the house wiring system if an ordinary socket is used.

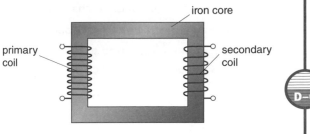

A simple transformer

How transformers work

- The changing **current** in the primary coil produces a changing **magnetic field** in the core.
- The changing magnetic field in the core induces a changing voltage in the secondary coil.

$$\frac{\text{voltage across the primary } (V_p)}{\text{voltage across the secondary } (V_s)} = \frac{\text{number of turns on primary coil } (N_p)}{\text{number of turns on secondary coil } (N_s)}$$

- The isolating transformer has the same number of turns on both secondary and primary coils.
  - The mains supply is hidden. The output terminals are not live so there is no danger of electrocution if you touch them with wet hands.

Energy loss

- When a current passes through a wire, the wire gets hot.
- The transformer coils and the overhead **power** lines of the **National Grid** get hot and lose **energy** to the surroundings.

Reducing transmission loss

- Power is a measure of how fast energy is transferred.

  power loss – (current² × **resistance**)

- The electrical power supplied to the primary coil of a transformer depends on the input voltage and current.
- Input power is given by:

  $P_p = V_p I_p$

- Similarly, the output power of the secondary coil is given by:

  $P_s = V_s I_s$

- If the transformer is 100% efficient, input power = output power i.e. $V_p I_p = V_s I_s$
- In a step-up transformer, increasing the voltage leads to a decrease in current by the same factor. This assumes that the transformer is 100% efficient.
  - For example, the step-up transformer at the power station increases the voltage by a factor of 16 to 400 kV. This reduces the current in the overhead cables to $\frac{1}{16}$ of what it would be if electricity was transmitted at 25 000 V.

Improve your grade

Transformers

Tammy is using a transformer to give a 12 V supply to her laptop computer from the mains voltage of 240 V. The input coil has 400 turns on it.

(a) How many turns are on the output coil?

(b) i What is the output current if the input current is 2 mA?

   ii What assumption have you made in calculating your answer? *AO2* [3 marks]

Charging

Diode characteristics

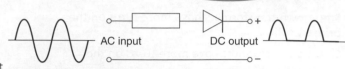

- Use the same circuit for **diode** characteristics as you would for a **resistor** or a bulb.
- Most diodes need a **voltage** of about 0.6 V before they start to work.
- When a diode is connected so that a **current** passes, it is forward biased.
- When a diode is connected so that no current passes, it is reverse biased.
- An AC supply to a single diode produces a half-wave rectified output.
- A **resistor** is always used in series with a diode to protect the diode.

Remember!
Remember to reverse the **power** supply when examining diode characteristics.

AC input   DC output

Rectifier circuit

How a diode works

- A diode consists of a piece of **n-type** and a piece of **p-type** semi-conductor joined together.
- N-type semi-conductor has an excess of **electrons**.
- P-type semi-conductor has a shortage of electrons – the gaps are called holes.
- The space either side of the junction has no electrons or holes.
- If the positive terminal of the supply voltage is connected to the n-type semi-conductor, the space widens and no current passes.
- If the positive terminal of the supply voltage is connected to the p-type semi-conductor, the space narrows, eventually disappears and a current passes.

n-type semi-conductor has an excess of electrons

p-type semi-conductor has a shortage of electrons – the 'gaps' are called **holes**

key:
+ positive ion
− negative ion
● electron
✳ 'hole'

n-type and p-type semi-conductors

Rectifier circuits

- Four diodes can be arranged to make a **bridge circuit**.
- The addition of a large **capacitor** makes the output smoother.

Capacitor action

- When a DC supply is connected to a capacitor, the capacitor becomes charged. The voltage across the capacitor increases until it is equal to the supply voltage.
- When the capacitor is connected to a resistor, for example, the voltage decreases as the capacitor discharges.

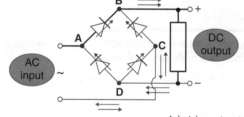

AC input ~

DC output

A bridge circuit

Rectification

- In a bridge circuit, during the positive half-cycle, the current passes from A to B, to the external circuit, to D then C and back to the AC supply. During the negative half-cycle, the current passes from C to B, to the external circuit, to D then A and back to the AC supply.
  – This means that the DC output is always positive at B and negative at D.

Storing charge

- The chemicals in a **battery** continue to produce **energy** for a battery to use. When the chemicals are used up, the battery is no longer useful.
- The capacitor stores electrical energy. There is no continual energy source.

Smoothing

- A smoothing capacitor acts as a reservoir. When the DC voltage from the rectifier circuit falls, the capacitor supplies current to the output. The capacitor charges near the peak value of the varying DC.

Remember!
The time it takes for a capacitor to discharge depends on the resistance and the capacitor.

Improve your grade

Rectification
Explain how a bridge circuit can turn AC supply into a DC output. *AO1* [3 marks]

P6 Summary

A potential divider consists of two resistors used to produce a specific voltage.

A variable resistor is used to set the threshold level for an output voltage.

When resistors are connected in series the total resistance is
$R_t = R_1 + R_2 + R_3$
and when connected in parallel
$$\frac{1}{R_t} = \frac{1}{R_1} + \frac{1}{R_2} + \frac{1}{R_3}$$

A diode only allows a current to pass in one direction.

Diodes are made from n-type and p-type semi-conductors joined together.

The resistance of a light dependent resistor (LDR) decreases as the light intensity increases.

The resistance of a thermistor decreases as the temperature increases.

LDRs and thermistors can be used as one of the resistors in a potential divider.

Electric circuits

Diodes can be arranged as a rectifier to change AC into DC.

Resistors control the current in a circuit.

resistance = voltage ÷ current

The resistance of a wire increases as the wire gets hotter.

Longer wires have more resistance than shorter wires.

The resistance of an ohmic conductor is found from the gradient of the voltage against current graph.

A capacitor stores electric charge.

A capacitor smoothes the output from a rectifier.

The behaviour of NOT, AND, OR, NAND and NOR gates can be described using truth tables.

Inputs and outputs are described as high (1) and low (0).

Logic gates work at low voltages (about 5–6 volts).

Inputs to logic gates can be controlled by potential divider circuits.

Logic circuits

A relay uses a small current to switch a larger current.

The output from a logic gate is a small current and a relay is often used to allow this to switch on a motor, heater etc.

In a transistor a small base current can switch on a greater current through the collector.

$I_e = I_b + I_c$

Two transistors can be combined to make an AND gate.

Other logic gates can be made with different combinations of transistors.

Transistor circuits always have a high value resistor in the base circuit to limit the current.

A current carrying wire has a magnetic field around it.

A motor moves because of the interaction between the magnetic field due to a current carrying coil and a permanent magnet.

Fleming's left hand rule allows you to predict the direction in which a motor turns.

A commutator makes sure the motor keeps spinning in the same direction.

Motors spin faster when:
- the magnetic field is stronger
- the current is greater or
- there are more turns on the coil.

In a generator a current is induced in a coil of wire when there is a changing magnetic field.

The quicker the field changes, the greater the current.

Slip rings allow AC to be generated.

Brushes are contacts that touch the slip rings and complete the circuit.

Electromagnetism

A transformer changes the size of an AC voltage.

$$\frac{V_p}{V_s} = \frac{N_p}{N_s}$$

If the transformer is 100% efficient

power input = power output

$V_p I_p = V_s I_s$

Electricity is transmitted at high voltages to reduce the current and hence reduce energy loss in the overhead power lines.

power loss = current² × resistance

Page 4 Specific heat capacity

Ed uses a stainless steel saucepan to heat his soup from 17 °C to 94 °C. The saucepan has a mass of 1.1 kg and a specific heat capacity of 510 J/kg °C. Energy is required to heat the soup.

(a) Calculate the extra energy required to raise the temperature of the saucepan. *AO2* [2 marks]

(b) Ed reads that a 1.1 kg copper saucepan will be more energy efficient. Explain why. *AO2* [2 marks]

(a) energy = 1.1 × 510 × 94 = 52 734 J

(b) Copper is a better conductor of heat than stainless steel so less energy will be lost in heating the saucepan.

Answer grade: D–C. In (a) the final temperature has been used instead of the temperature change. The student should show their workings as this might gain some marks. In (b) the statement is correct but does not answer the question. For full marks a description of the difference in the specific heat capacities is needed.

Page 5 Energy loss in a cavity wall

The Johnson's house has cavity walls. They decide to have foam injected into the cavity to reduce energy loss.

Explain how energy is transferred to the roof space from the cavity. *AO1* [3 marks]

The air in the cavity moves into the roof space by convection. The foam traps the air so it cannot move.

Answer grade: D–C. Sentence 1 states, but does not explain, the process. Sentence 2 explains the reason why foam is used but does not answer the question. For full marks, the student should include a description of convection as being due to the changing density of air as it is warmed.

Page 6 Diffraction effects

Light is diffracted as it passes through a narrow slit. Describe how the amount of diffraction depends on the wavelength of the light and the width of the slit. *AO1* [2 marks]

The wider the slit, the less the diffraction.

Answer grade: C–B. The sentence is correct but it is not a complete answer. For full marks, the student should include the fact that maximum diffraction happens when the slit width and wavelength are a similar size.

Page 7 Sending signals

Adam is standing on top of a hill in line of sight and 10 km away from Becky who is on top of another hill. They can communicate either by using light, radio or electrical signals. Suggest one advantage and one disadvantage of using each type of signal. *AO1* [3 marks]

● *Light signals need a code*
● *Electrical signals need wires*
● *Radio signals can be intercepted.*

Answer grade: D–C. Bullet points can sometimes help to organise ideas. Disadvantages are given but no advantage. For full marks, the student should include a unique advantage and disadvantage.

Page 8 Microwave transmitters

The Telecom Tower in London is one of the tallest buildings in the city. There are many microwave aerials surrounding the top of the tower.

Explain why they are sited so high up. *AO1* [2 marks]

Microwaves do not diffract very much around buildings.

Answer grade: D–C. The sentence is correct but not a complete answer. For full marks, the student should include the fact that aerials need to be in line of sight.

Page 9 Advantages of using digital signals and optical fibres

Explain the advantages of using digital signals and optical fibres compared with analogue signals and electrical cables for data transmission. *AO1* [4 marks]

Analogue signals can show a lot of interference. Digital signals do not exhibit any interference. You can only send one signal down copper wires but an optical fibre allows multiplexing.

Answer grade: D–C. Sentence 1 states a reason for not using analogue signals. Sentence 2 does not give enough information. For full marks, the student should include reference to digital signals not showing interference because they only have 2 values. Sentence 3 is not totally true. It is possible to multiplex along copper wires, but less so than through optical fibres. For full marks, the student should explain the meaning of multiplexing.

Page 10 Radio communication

The picture shows a transmitter and receiver on the Earth's surface, out of line of sight.

(a) Explain how long wave radio signals travel from the transmitter to the receiver. *AO1* [3 marks]

(b) Explain how microwave signals travel from the transmitter to the receiver. *AO1* [2 marks]

(a) Long wave signals are reflected from the ionosphere, then from the sea, then the ionosphere again and so on until they reach the receiver.

(b) Microwave signals have too high a frequency to be reflected back by the atmosphere. They are received and amplified by a satellite before being reflected back to Earth.

Answer grade: C–B. In (a), reflection from the sea is mentioned as well as from the ionosphere. For full marks, the student should refer to refraction and total internal reflection from the atmosphere. In (b), sentence 1 is irrelevant to the question. Sentence 2 describes microwaves being received but they are not reflected. For full marks, the student should refer to the satellite retransmitting them.

Page 11 Earthquake waves

An Earthquake occurs with its epicentre at **E**. It is detected at two monitoring stations **A** and **B**.

Describe and explain the appearance of the seismograph traces at **A** and **B**. *AO1/AO2* [4 marks]

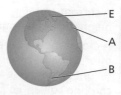

The trace at A will show a P wave and an S wave. The one at B will only show a P wave. S waves will not travel through the Earth's core so are not received at station B.

Answer grade: D–C. Sentences 1 and 2 describe the traces that will be seen. For full marks, the student should include the fact that the P wave is received before the S wave at A and that the P wave is received at B later than at A. Sentence 3 states why the wave is not received but does not explain why. For full marks, the student should explain that the core contains liquid.

Page 13 Photocells

A photocell contains two pieces of silicon joined together to make a p-n junction.

Explain how light falling on a photocell produces an electric current. *AO1* [3 marks]

n-type silicon has extra electrons which move to produce current.

Answer grade: D–C. The explanation of the p-n junction is incomplete. The answer can be improved by adding the description of p-type silicon as having an impurity added which leads to an absence of electrons and then explaining that energy from the Sun's photons allows the electrons to move across from n-type to p-type silicon.

Page 14 Generators

Describe three ways to increase the output of an electrical dynamo. *AO1* [3 marks]

Longer wire, bigger magnets and faster movement.

Answer grade: D–C. The phrase 'bigger magnets' is too vague and longer wire is incorrect. To improve this answer, add the use of a stronger magnetic field and more turns on the coil.

Page 15 The greenhouse effect

Describe the greenhouse effect and explain how it contributes to global warming. *AO1 [4 marks]*

The electromagnetic radiation from the Sun is absorbed by and warms the Earth. The Earth then re-radiates the energy which warms the atmosphere.

Answer grade: D–C. The description of what is happening is correct but the explanation of why this takes place is missing. For full marks, the student should include the idea that the Sun's radiation is at a shorter wavelength than that re-radiated by the Earth and also mention that the greenhouse gases absorb this longer wavelength radiation and this warms the atmosphere.

Page 16 Cost of electricity

Tracey's flat has electric storage heaters which heat up at night and release the heat slowly during the day. Why is this cheaper for Tracey? *AO1* [2 marks]

She pays less for electricity during the night.

Answer grade: D–C. This would gain one mark for a correct statement but the question is worth 2 marks, so a fuller explanation is required. For full marks, the student should include the fact that electrical power stations will not be switched off but will always be producing power and when the demand is lower at night the electricity will be sold more cheaply.

Page 17 Ionising radiation

Give two advantages and two disadvantages of nuclear power. *AO1* [4 marks]

Nuclear power does not pollute the environment and it is renewable
Nuclear power is expensive and risky.

Answer grade: D–C. Answer lacks detail and is vague. For full marks, the student should include clear advantages – nuclear power does not produce greenhouse gases; and disadvantages – costs of maintenance are high and there is a risk of nuclear accidents. Nuclear power is a non-renewable energy source.

Page 18 Unmanned space travel

Why do we send unmanned space probes to explore our Solar System? *AO1* [3 marks]

We explore our Solar System to find out more about the other planets and to search for alien life.

Answer grade: D–C. Whilst this answer may be true it is not addressing the question specifically focusing on unmanned space exploration. A better answer would have explained the advantages of not having humans on board, e.g. no risk to life, no food water or oxygen needed, lower costs, longer journeys possible, no need to return.

Page 19 Asteroids

What evidence have scientists gathered to show how the Moon was formed? *AO1* [3 marks]

The Moon was formed when there was a collision between the Earth and another planet. During the collision, iron became concentrated in the Earth's core.

Answer grade: D–C. This answer correctly describes how the Moon was formed. However, the question is asking for scientific evidence. For full marks, the student should include the fact that there is no iron on the Moon as it all merged with the Earth's core and the fact that the Moon has the same oxygen composition as the Earth.

Page 20 The Big Bang

What is red shift and how does it provide evidence for the Big Bang? *AO1* [4 marks]

Red shift is when the light from stars is shifted to the red end of the spectrum and it shows galaxies are expanding.

Answer grade: D–C. This gives an idea about red shift but the answer is not clearly explained. For full marks, the student should explain that red shift is when light from distant galaxies contains a spectrum of lines that are shifted towards a longer wavelength. This happens when the object emitting the light is moving away from the observer. Since light from most distant galaxies exhibits red shift and the further away they are the greater the red shift, this indicates that all galaxies are moving apart which is evidence for the Big Bang.

Page 22 Calculating time

How many hours will it take to travel 560 km at an average speed of 25 m/s? *AO1* [1 mark] *AO2* [2 marks]

$$Time = \frac{distance}{speed} = \frac{560}{25 \times 60 \times 60} = 6.2 \times 10^{-3} \, h$$

Answer grade: D–C. The method for the calculation is correct and the conversion from seconds to hours completed. For full marks it is important to check that all units are consistent. Metres should be converted to kilometres.

Page 23 Calculating acceleration

A car is travelling at 10 m/s. It accelerates at 4 m/s² for 8 s. How fast is it then going? *AO1* [1 mark] *AO2* [2 marks]

change in speed = acceleration × time = 4 × 8 = 32 m/s

Answer grade: D–C. The working for the change in speed is correct but this needs to be added to the original speed of 10 m/s. For full marks, it is important to read the question carefully and double check the calculations.

Page 24 Stopping distance

Explain why brakes and tyres are checked when a car has its annual MOT Test. *AO2* [2 marks]

If they are worn, the braking distance increases.

Answer grade: D–C. The sentence is correct but does not provide a full explanation. For full marks, the answer should explain that worn brakes and worn tyres reduce friction and hence the braking force.

Page 25 Calculating driving force

A Ford Focus has a power rating of 104 kW.

(a) Calculate the resultant force acting when the car is travelling at 90 km/h.

(b) Explain how this force compares with the driving force of the engine. *AO1* [3 marks] *AO2* [1 mark]

(a) $force = \dfrac{power}{speed} = \dfrac{104 \times 1000 \times 3600}{90 \times 1000} = 4160 \, N$

(b) *It would be larger.*

Answer grade: D–C. The calculation is correct and the working is clearly shown. The answer to part (b) is ambiguous and is only a description, not an explanation. For full marks, the answer should avoid the use of the word 'it' and explain that the driving force is larger because there is drag and other frictional forces to overcome.

Page 26 Carbon dioxide emissions

Some scientists suggest that carbon dioxide emissions from burning bio-fuels may be at least 20% lower than those from fossil fuels. Some scientists argue that overall the emissions may be higher than from fossil fuels. Suggest why emissions may be higher. *AO3* [3 marks]

Deforestation and processing the bio-fuel.

Answer grade: D–C. The sentence gives two correct reasons but not full explanations. For full marks, the answer should explain that deforestation leads to an increase in carbon dioxide levels and electricity is needed in the processing of the bio-fuel, which is generated in a power station and releases carbon dioxide.

Page 27 Seat belts

Some people think that wearing a seat belt should be up to the individual, not the law. Explain how the wearing of seat belts can help to avoid injury but may not always do so. *AO1* [3 marks]

Seat belts stretch and absorb energy in a crash. They increase the time it takes for a person to come to rest therefore reducing the force since

$$force = \frac{change \ in \ momentum}{time}$$

Answer grade: C–B. Both sentences are correct and are a good answer to the first part of the question. For full marks, the answer should explain the circumstances where a seat belt may be a disadvantage, e.g. trapping the person in a burning or submerged car.

Page 28 Gravitational field strength

(a) Explain why a 1 kg ball, dropped from a height of 10 m above the ground at the North Pole, takes less time to reach the ground than an identical 1 kg ball dropped from 10 m above the ground at the equator.

(b) How would the time taken be different if the ball was taken to the top of a mountain and dropped from a height of 10 m? *AO1* [2 marks], *AO2* [1 mark]

(a) *The acceleration due to gravity is greater at the poles than it is at the equator.*

(b) *The acceleration due to gravity is less so it takes less time to fall.*

Answer grade: C–B. The answer to part (a) is correct but does not give the reason why the acceleration is greater. The first sentence in part (b) is correct but the second contradicts the first. For full marks, it should explain that the acceleration at the poles is greater because the Earth is not a perfect sphere and the poles are nearer the centre. Answers should always be read carefully to make sure there are no contradictions. If acceleration is less, the time will be greater.

Page 29 Energy changes during free fall

Mel is a free-fall parachutist. During her time in free fall she reaches terminal speed. Explain how her gravitational potential energy and kinetic energy change during her descent. *AO1* [3 marks]

As she falls, her gravitational potential energy decreases as it is transferred into kinetic energy. When she reaches terminal speed, her kinetic energy remains constant.

Answer grade: C–B. The first sentence implies that KE increases as she falls and is worthy of credit. The second sentence does not explain what happens to the GPE if it is not transferred into KE. For full marks, the answer should explain that as she leaves the plane, her KE is zero, GPE is a maximum and that this reduces as the KE increases. It should also explain that the GPE is used to do work against friction, resulting in the heating of the air.

Page 31 Static charge

Connor is in the library walking on a nylon carpet.
He touches a metal bookshelf and receives an electric shock.
Explain how he became charged and why he received a
shock. *AO2* [3 marks]

*He becomes charged and when he touches the shelf he
discharges.*

Answer grade: D–C. The basic ideas are correct but there is
insufficient detail in the answer. The answer should explain
that it is the friction of walking on the nylon carpet that
leads to the transfer of electrons charging him up. Then the
charges flow to earth through the metal shelf causing a
shock.

Page 32 Spray painting

Static electricity is useful in spray-painting cars.

Explain how by writing about:
- electrostatic charge
- electrostatic force
- why it is used. *AO1* [3 marks]

*The paint experiences an electrostatic force because it is
electrostatically charged. Paint will stick to the car better.*

Answer grade: D–C. The answer is very brief and lacks
detail. Charge on the plates is induced by the dust particles
which are then attracted because opposite charges attract.

The answer simply repeats the words in the question
without showing understanding. For full marks, it should
explain that the spray gun is charged so that this in turn
charges the paint and that the car is charged oppositely to
the paint so that the paint is attracted and sticks to the car,
and less is wasted.

Page 33 Electrical safety

Explain how the fuse and earth wire operate to protect a
user. *AO1* [2 marks]

If a fault occurs the fuse blows stopping the current.

Answer grade: D–C. The answer is correct but does not
include an explanation of the role of the earth wire. For full
marks, the answer should explain that a fault would cause a
large current to flow through the earth wire causing the
fuse to overheat and melt.

Page 34 Ultrasound and imaging

(a) State one similarity and one difference between sound
and ultrasound waves.
(b) Give two reasons why doctors may decide to use
ultrasound instead of x-rays to get images of inside
the body.

AO1 [4 marks]

(a) *Both are waves but ultrasound is at a much higher
frequency.*
(b) *Ultrasound is not harmful and it can make images.*

Answer grade: D–C. (a) the fact that these are waves is
given in the question. They are both longitudinal waves
would be a better answer. It is also a good idea to add the
fact that ultrasound is above 20 000 Hz.

X-rays are an ionising radiation and this is why they could
be harmful. Both can produce images but the ultrasound
shows soft tissue better.

Page 35 Nuclear radiation

When carbon-14 undergoes beta decay what happens to the
atom and what new element is formed? *AO1* [2 marks]

*In the nucleus of the atom a neutron changes into a proton
and an electron is emitted at high speed. The carbon atom is
now carbon-13.*

Answer grade: C–B. The first part of the answer is correct
and quite detailed. In the second part a mistake has been
made – the atomic number, not the mass number, should be
changed so a new element is formed nitrogen-14.

Page 36 Radioisotope dating

Carbon-14 has a half-life of 5700 years.

(a) What is meant by the half-life of a radioactive sample?
(b) A sample of bone was found to have a 25% of the
amount of carbon-14 found in a living organism. How old
was this bone?

AO1 [1 mark], *AO2* [2 marks]

(a) *The half-life of a radioactive sample is the time taken for
half of the nuclei to decay.*
(b) *The sample was 25% of 5 700 = 1425 years old.*

Answer grade: D–C. The answer to part (a) is correct but
should say the average time taken. In (b) it has not been
understood that 25% is one quarter, which means the
amount of carbon-14 nuclei would have halved and then
halved again. So 2 half-lives must have passed or 11 400 years.

Page 37 Medical tracers

Explain what a medical tracer is and how suitable materials
are chosen to be used as tracers. *AO1* [3 marks]

*A tracer is a substance put into the body to find out if there is
a problem. It must be able to be detected outside of the body
and must not harm the patient.*

Answer grade: D–C. The answer gives two correct statements
but not full explanations. For full marks, the answer should
explain that a tracer is a radioisotope that will emit radiation
that can be detected from the outside of the body. It should
also explain that since it will be emitting potentially harmful
radiation it must have a short half-life so that this is limited,
and finally that a gamma emitter is used as it is the most
penetrating and the least ionising radiation.

Page 38 Controlling nuclear fission

Explain how the output of a nuclear reactor is controlled.
AO1 [3 marks]

*The graphite moderator slows down the neutrons to make
them more likely to be captured by uranium. Boron rods
absorb neutrons to slow down the rate of fission.*

Answer grade: C–B. This is a good answer giving the two
ways the chain reaction is controlled. For full marks, the
answer should explain that the boron control rods can be
raised or lowered to absorb different numbers of neutrons
when required.

Page 40 Satellite speed

Calculate the approximate speed of a satellite in low polar orbit. Assume the radius of the Earth is 6400 km, the altitude of the satellite 100 km and the time taken to orbit 90 minutes. *AO1* [1 mark] *AO2* [2 marks]

distance $= 2\pi r = 40840$ *km*

$$speed = \frac{distance}{time} = \frac{40840}{90} = 454 \ km/h$$

Answer grade: D–C. The method for the calculation is correct but the time has not been changed into hours. Always show how you work out an answer because you will often gain some marks. For full marks, always check that the units are correct.

Page 41 Scalars and vectors

(a) Explain the difference between a scalar and a vector. *AO1* [1 mark]
(b) Which of the following are vectors and which are scalars?

mass   temperature   force   acceleration   volume
AO1 [2 marks]

(a) A vector has both size and direction.
(b) Mass and temperature are scalars, so is volume

Answer grade: D. There is nothing inaccurate in the answers, but they are both incomplete and only score one mark as a result. To score the mark for part (a), state that a scalar only has size. To score both marks for part (b) state that force and acceleration are vectors.

Page 42 Projectiles

Explain why a satellite in orbit can be thought of as being like a projectile. *AO2* [2 marks]

The harder and faster you throw something from the top of a mountain, the further it goes before it hits the ground. If you throw it hard enough, it goes into orbit.

Answer grade: D–C. Although both statements are true, they are lacking in detail. A full explanation should include the shape of the trajectory as a parabola, the object is attracted towards the Earth by gravity and an acknowledgement that as the object falls towards the Earth, the surface of the Earth is curving away from it.

Page 43 Explosions

When a radioactive nucleus emits a beta particle, the nucleus moves slowly in the opposite direction to the fast moving beta particle. Explain why. *AO2* [2 marks]

The nucleus is much heavier than the beta particle so moves much more slowly.

Answer grade: D–C. The statement is correct but for full marks an explanation of conservation of momentum is needed. The initial momentum is zero so when the beta particle is emitted in one direction the nucleus must move in the opposite direction.

Page 44 Geostationary satellites

Geostationary satellites are used for communication. Explain why the transmitting and receiving dishes need very careful alignment and why the satellites must be as far away from each other as possible. *AO2* [3 marks]

The size of the dish is very large which means there is little diffraction. The beam is very focused so the dishes must be lined up. If they are too close together, signals may overlap.

Answer grade: D–C. The first sentence is correct but would be better if there were a comparison between the dish size and the microwave wavelength. There is clear mention of diffraction. The beam is very narrow as well as focused and this should be mentioned. The final sentence does not make it clear that it is the closeness of the satellites that is being referred to. Words such as it or they should be avoided in answers. If the satellites are too close, one may receive a signal meant for another.

Page 45 Interference

An oscillator, connected to two loudspeakers, is producing a sound with a wavelength of 0.4 m. Jane stands 5.5 m away from one loudspeaker and 7.7 m away from the other. Explain what she will hear. *AO2* [4 marks]

The path difference is 2.2 m so there will be destructive interference.

Answer grade: D–C. The statement is correct but is not a full explanation. For full marks, the fact that a half wavelength is 0.2 m, so 2.2 m is an odd number of half wavelengths, should be noted. The answer should also conclude that destructive interference would result in a quiet/silent note being heard.

Page 46 Dispersion

Explain why white light passing into a prism leaves a spectrum of colours. *AO1* [2 marks]

The colours are refracted by different amounts.

Answer grade: E–D. The statement is really only a rewording of the question – not an explanation. For full marks students should explain that the speed of light is different for different colours and that this leads to different refractive indices.

Page 47 Camera lenses

A simple camera has a fixed lens.
(a) Draw a ray diagram to show where the image is formed if the object is a long way away from the camera.
(b) How far from the lens is the film in the camera?
AO2 [3 marks]

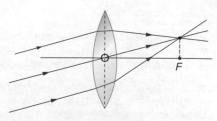

The distance is 2 cm.

Answer grade: D. The diagram is clearly and accurately drawn but for full marks it needs to include labels showing the image and the focal point or focal plane. Although the distance measured is correct, since there were no numbers quoted in the question, the answer should be general in terms of focal length.

Page 49 Resistance

Susan is using a 1 m long piece of nichrome wire to investigate the effect of length on resistance in the science laboratory. The teacher has chosen a piece of wire with a resistance of 100 Ω.

(a) Calculate the current passing when Susan puts 2 V across the ends of the wire.

(b) Sketch the graph you would expect Susan to have at the end of her investigation. AO2 [3 marks]

(a) 0.02

(b)

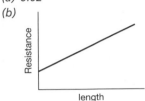

Resistance is proportional to length so straight line.

Answer grade: D–C. The answer for the calculation is correct but no workings have been shown and no units given. Workings should always be shown because these may gain some marks. For full marks, students should check that the units are correct. The statement in part (b) is correct but the graph does not show a directly proportional relationship as it does not pass through the origin (0,0). Students should always check to make sure there are no contradictions in their answers.

Page 50 Resistors in parallel

Calculate the total resistance when three 100 Ω resistors are connected in parallel. AO2 [2 marks]

$$\frac{1}{R} + \frac{1}{R} + \frac{1}{R} = \frac{1}{100} + \frac{1}{100} + \frac{1}{100} = \frac{3}{100} = 0.03$$

Answer grade: C–B. This is a good attempt and the student has shown workings so marks can be gained for this.

The problem is that the whole formula has not been written out and the final answer is

$$\frac{1}{R_{tot}} = 0.03, \text{ so } R_{tot} = \frac{1}{0.03} = 33.3 \text{ Ω}.$$

Page 51 Logic gates

Draw a circuit diagram to show how an AND gate can be constructed from two transistors. AO1 [3 marks]

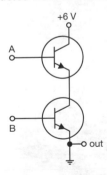

Answer grade: C–B. A very clear diagram but there should be resistors in both of the inputs as transistors in logic gates work on very small input currents.

Page 52 Truth tables

Tom builds a logic circuit with three logic gates as shown right. Draw a truth table for Tom's circuit. AO2 [2 marks]

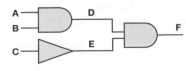

A	B	C	D	E	F
1	0	0	0	1	0
0	1	0	0	1	0
0	0	1	0	0	0
1	1	1	1	0	0
0	0	0	0	1	0
1	1	0	1	1	1

Answer grade: D–C. The truth table is correct but for full marks it must be complete with *all* possible inputs considered. It is best to set out the inputs in a logical manner.

Page 53 DC electric motor

Jasmine has made a model electric motor. What three things can she change to make her motor spin faster? AO1 [3 marks]

Greater current, bigger magnets and more coils.

Answer grade: D–C. The first point is correct but the phrase 'bigger magnets' is not clear - the magnetic field needs to be stronger would be a better statement. More 'coils' could also be made clearer by saying more turns on the coil. Answers should always be very clear.

Page 54 Generators

Explain how an AC generator works including the action of the slip rings and brushes. AO1 [4 marks]

As the coil rotates induced current will always be going the same way. An alternating current is induced. Slip rings are connected to the ends of the coil to allow the coil to spin without the wire being tangled.

Answer grade: D–C. The statements are correct but are not a full explanation. For full marks, students should explain that the coil flips over each half turn but that the current is always in the same direction.

Page 55 Transformers

Tammy is using a transformer to give a 12 V supply to her laptop computer from the mains voltage of 240 V. The input coil has 400 turns on it.

(a) How many turns are on the output coil?

(b) (i) What is the output current if the input current is 2 mA?

   (ii) What assumption have you made in calculating your answer? AO2 [2 marks]

(a) 20 turns   (b)(i) 40 mA   (ii) assuming it is AC.

Answer grade: C–B. The calculations are correct but show no workings. Students should always show workings as partial credit will be given even if your answer is incorrect. The last part of the answer is incorrect – the assumption that has to be made is that the transformer is 100% efficient and no power is lost.

Page 56 Rectification

Explain how a bridge circuit can turn AC supply into a DC output. AO1 [3 marks]

During one cycle current passes to A, then B, from the DC output to the circuit, back to D, then C and finally back to the AC supply.

During the other cycle current passes to C, then B, from the DC output to the circuit, back to D, then A and finally back to the AC supply.

Answer grade: B. The explanation is good and makes use of the letters on the diagram to help with the explanation. For full marks it is important that students note that each part happens in a *half* cycle not a whole cycle.

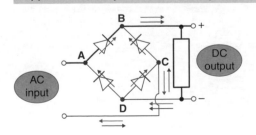

Understanding the scientific process

As part of your Physics assessment, you will need to show that you have an understanding of the scientific process – How Science Works.

This involves examining how scientific data is collected and analysed. You will need to evaluate the data by providing evidence to test ideas and develop theories. Some explanations are developed using scientific theories, models and ideas. You should be aware that there are some questions that science cannot answer and some that science cannot address.

Collecting and evaluating data

You should be able to devise a plan that will answer a scientific question or solve a scientific problem. In doing so, you will need to collect data from both primary and secondary sources. Primary data will come from your own findings – often from an experimental procedure or investigation. While working with primary data, you will need to show that you can work safely and accurately, not only on your own but also with others.

Secondary data is found by research, often using ICT – but do not forget books, journals, magazines and newspapers are also sources. The data you collect will need to be evaluated for its validity and reliability as evidence.

Presenting information

You should be able to present your information in an appropriate, scientific manner. This may involve the use of mathematical language as well as using the correct scientific terminology and conventions. You should be able to develop an argument and come to a conclusion based on the recall and analysis of scientific information. It is important to use both quantitative and qualitative arguments.

Changing ideas and explanations

Many of today's scientific and technological developments have both benefits and risks. The decisions that scientists make will almost certainly raise ethical, environmental, social or economic questions. Scientific ideas and explanations change as time passes and the standards and values of society change. It is the job of scientists to validate these changing ideas.

How science ideas change

From the information you have learnt, you will know that science is a process of developing, then testing, theories and models. Scientists have been carrying out this work for many centuries and it is the results of their ideas and trials that have provided us with the knowledge we have today.

However, in the process of developing this knowledge, many ideas were put forward that seem quite absurd to us today.

In 1692, the British astronomer Edmund Halley (after whom Halley's Comet was named) suggested that the Earth consisted of four concentric spheres. He was trying to explain the magnetic field that surrounds the Earth and suggested that there was a shell of about 500 miles thick, two inner concentric shells and an inner core. Halley believed that these shells were separated by atmospheres, and each shell had magnetic poles with the spheres rotating at different speeds. The theory was an attempt to explain why unusual compass readings occurred. He also believed that each of these inner spheres, which was constantly lit by a luminous atmosphere, supported life.

Reliability of information

It is important to be able to spot when data or information is presented accurately. Just because you see something online or in a newspaper, does not mean that it is accurate or true.

Think about what is wrong in this example from an online shopping catalogue. Look at the answer at the bottom of the page to check that your observations are correct.

From box to air in under two minutes!

Simply unroll the airship and, as the black surface attracts heat, watch it magically inflate.

Seal one end with the cord provided and fly your 8-metre, sausage-shaped kite.

- Good for all year round use.

- Folds away into box provided.

- A unique product – not for the faint hearted.

- Educational as well as fun!

Once the airship is filled with air, it is warmed by the heat of the sun.

The warm air inside the airship makes it float, like a full-sized hot-air balloon.

Answer
Black absorbs heat, it does not attract it.

Glossary

A

ABS braking system, known as advance braking system, which helps to control a skidding car

absorbed taken in

acceleration a measurement of how quickly the speed of a moving object changes (when speed is in m/s the acceleration is in m/s^2)

action (force) applying a force on an object

aerial a device for receiving or transmitting radio signals

air bags cushions which inflate with gas to protect people in a vehicle accident

air resistance the force exerted by air to any object passing through it

alpha particles radioactive particles which are helium nuclei – helium atoms without the electrons (they have a positive charge)

alternating current an electric current that is not a one-way flow

amplitude the amplitude of a wave is the maximum displacement of a wave from its rest position

analogue signal a signal that shows a complete range of frequencies; sound is analogue

AND gate a logic gate which only delivers an output if both input terminals are on

angle of incidence the angle between the incident ray of light and the normal at a given point

angle of refraction the angle between the refracted ray of light and the normal at a given point

aperture the size of the hole through which light enters a camera

asteroid composed of rock or metallic material orbiting the Sun in a region between Mars and Jupiter

atmosphere mixture of gases above the lithosphere, mainly nitrogen and oxygen

atom the basic 'building block' of an element which cannot be chemically broken down

atomic number the number of protons found in the nucleus of an atom

attenuate lose energy

attract move towards, for example, unlike charges attract

average speed total distance travelled divided by the total time taken for a journey

B

background radiation ionising radiation from space and rocks, especially granite, that is around us all the time but is at a very low level

bacteria single-celled micro-organisms which can either be free-living organisms or parasites (they sometimes invade the body and cause disease)

battery two or more electrical cells joined together

beta particles particles given off by some radioactive materials (they have a negative charge)

Big Bang the event believed by many scientists to have been the start of the universe

biofuels fuels made from plants – these can be burned in power stations

black hole a region of space from which nothing, not even light, can escape

boron control rods rods that are raised or lowered in a nuclear reactor to control the rate of fission

braking distance distance travelled while a car is braking

C

cancer life-threatening condition where body cells divide uncontrollably

camera an optical instrument that produces a reduced image on a piece of film (film camera) or light sensitive chip (digital camera)

capacitor an electronic device for storing electric charge

carbon dioxide (CO_2) gas present in the atmosphere at a low percentage but important in respiration, photosynthesis and combustion; a greenhouse gas which is emitted into the atmosphere as a by-product of combustion

carbon-14 radioactive isotope of carbon

cement the substance made when limestone and clay are heated together

centripetal force force towards the centre of a circle essential for circular motion

CFCs gases which used to be used in refrigerators and which harm the ozone layer

chain reaction a reaction where the products cause the reaction to go further or faster, e.g. in nuclear fission

charge(s) a property of matter charge exists in two forms, positive and negative, which attract each other

circuit breakers resettable fuses

'cold fusion' attempts to produce fusion at normal room temperature that have not been validated since other scientists could not reproduce the results

comets lumps of rock and ice found in space – some orbit the Sun

commutator part of an electric motor that reverses the current direction every half turn

compressions particles pushed together, increasing pressure

computer modelling using a computer to 'model' situations to see how they are likely to work out if you do different things

condenser lens used in a projector to concentrate light on a slide

constructive (interference) when two waves combine to give increased amplitude

converging coming towards a point

convex curving outwards

core the centre part of a planet, made of iron

cosmic rays radiation from space that contributes to background radiation

count rate average number of nuclei that decay every second

crests peaks of a wave

critical angle the angle of incidence for which the angle of refraction is 90°; larger angles of incidence result in total internal reflection

crumple zones areas of a car that absorb the energy of a crash to protect the centre part of the vehicle

current flow of electrons in an electric circuit

curved line line of changing gradient

D

deceleration a measurement of how quickly the speed of a moving object decreases

defibrillator machine which gives the heart an electric shock to start it beating regularly

deforestation removal of large area of trees

Glossary

density the density of a substance is found by dividing its mass by its volume

destructive (interference) when two waves cancel to give reduced amplitude

diffraction the spreading out of a wave when it passes through a gap or around an edge

digital (signal) signal which is either 'on' or 'off'

diode an electronic component that only lets current pass through it in one direction

dispersion the splitting of light into its different wavelengths

distance–time graph a plot of the distance moved against the time taken for a journey

diverging spreading out / moving away from a point

DNA molecule found in all body cells in the nucleus – its sequence determines how our bodies are made (e.g. do we have straight or curly hair), and gives each one of us a unique genetic code

drag energy losses caused by the continual pushing of an object against the air or a liquid

dummies used in crash testing to learn what would happen to the occupants of a car in a crash

dynamo a device that converts energy in movement into energy in electricity

E

earth wire the third wire in a mains cable which connects the case of an appliance to the ground so that the case cannot become charged and cause an electric shock

earthed (electrically) connected to the ground (at 0 V)

echoes reflection of sound (or ultrasound)

efficiency ratio of useful energy output to total energy input; can be expressed as a percentage

electromagnet a magnet which is magnetic only when a current is switched on

electromagnetic spectrum electromagnetic waves ordered according to wavelength and frequency – ranging from radio waves to gamma rays

electromagnetic waves a group of waves that carry different amounts of energy – they range from low-frequency radio waves to high-frequency gamma rays

electron small negatively-charged particle

elliptical orbit a path that follows an ellipse – which looks a bit like a flattened circle

energy the ability to 'do work', for example the human body needs energy to function

enriched uranium uranium containing more of the U-235 isotope than occurs naturally

escape lane rough-surfaced uphill path adjacent to a steep downhill road enabling vehicles with braking problems to stop safely

exhaust gases gases discharged into the atmosphere as a result of combustion of fuels

explosion a very fast reaction making large volumes of gas

F

Fleming's left-hand rule if the fingers of the left hand are placed around a wire so that the thumb points in the direction of electron flow, the fingers point in the direction of the magnetic field produced by the conductor

focal length the distance from the optical centre of a lens to its focus

focal plane the plane that includes the focus (focal point)

focal point focus (of a lens)

focus (of lens or mirror) the point to which rays of light converge or from which they diverge

force a push or pull which is able to change the velocity or shape of a body

fossil fuels fuels such as coal, oil and gas

free-fall a body falling through the atmosphere without an open parachute

frequency the number of vibrations per second, frequency is measured in hertz

friction energy losses caused by two or more objects rubbing against each other

fuel consumption the distance travelled by a given amount of fuel, e.g. in km/100 litres

fuel rods rods of enriched uranium produced to provide fuel for nuclear power stations

fuse(s) a special component in an electric circuit containing a thin wire which is designed to melt if too much current flows through it, breaking the circuit

fusion the joining together of small nuclei, such as hydrogen isotopes, at very high temperatures with the release of energy

fusion bombs hydrogen bombs or H-bombs based on fusion reactions

G

gamma rays ionising electromagnetic waves that are radioactive and dangerous to human health – but useful in killing cancer cells

generator a device for converting energy of movement (kinetic energy) into electrical energy (current flow)

geostationary satellite a satellite in orbit above the equator taking 24 hours for one orbit

global warming the increase in the Earth's temperature due to increases in carbon dioxide levels

gradient rate of change of two quantities on a graph; change in y/change in x

granite an igneous rock containing low levels of uranium

graphite a type of carbon used as a moderator in a nuclear power station

gravitational attraction force of attraction between two bodies due to their mass

gravitational field strength the force of attraction between two masses

gravitational potential energy the energy a body has because of its position in a gravitational field, e.g. an object

gravity an attractive force between objects (dependent on their mass)

greenhouse gas any of the gases whose absorption of infrared radiation from the Earth's surface is responsible for the greenhouse effect, e.g. carbon dioxide, methane, water vapour

H

half-life average time taken for half the nuclei in a radioactive sample to decay

helium second element in periodic table; an alpha particle is a helium nucleus

hertz (Hz) units for measuring wave frequency

I

infrared waves non-ionising waves that produce heat – used in toasters and electric fires

Glossary

instantaneous speed the speed of a moving object at one particular moment

insulation a substance that reduces the movement of energy; heat insulation in the loft of a house slows down the movement of warmth to the cooler outside

interference waves interfere with each other when two waves of different frequencies occupy the same space; interference occurs in light and sound and can produce changes in intensity of the waves

iodine radioactive isotopes of iodine are used in diagnosing and treating thyroid cancer

ionisation the formation of ions (charged particles)

ionises adds or removes electrons from an atom, leaving it charged

ionosphere a region of the Earth's atmosphere where ionisation caused by incoming solar radiation affects the transmission of radio waves; it extends from 70 km to 400 km above the Earth's surface

ions charged particles (can be positive or negative)

isolating transformer a transformer whose output voltage is the same as its input voltage

isotopes atoms with the same number of protons but different numbers of neutrons

J

joule unit of work done and energy

K

kilowatt hour unit of electrical energy equal to 3 600 000 J

kinetic energy the energy that moving objects have

L

laser source of intense, narrow beam of light 'Light Amplification by Stimulated Emission of Radiation'

latent heat the energy needed to change the state of a substance

lead heaviest element having a stable isotope; all isotopes of the elements above it in the periodic table are unstable

lens a piece of transparent material, often glass, which is fatter in the middle than at the edges (convex) or thinner in the middle than at the edges (concave)

light-dependent resistor (LDR) device in an electric circuit whose resistance falls as the light falling on it increases

light-emitting diode (LED) a very small light in electric circuits that uses very little energy

light-year a unit of distance equal to the distance light travels through space in one year

linear a line of constant gradient on a graph

live (wire) carries a high voltage into and around the house

logic circuits circuits composed of a series of logic gates

logic gates electronic components that respond to signals by following preset logical rules

longitudinal (wave) wave in which the vibrations are in the same direction as the direction in which the wave travels

M

magnetic field an area where a magnetic force can be felt

magnification the ratio of the height of the image to the height of the object

magnitude size of something

mass number number of protons and neutrons in a nucleus

meteors bright flashes in the sky caused by rocks burning in the Earth's atmosphere

microwaves non-ionising waves used in satellite and mobile phone networks – also in microwave ovens

moderator material used to slow down neutrons in a nuclear power station

momentum the product of mass and velocity of an object. Unit: kgm/s or Ns

Morse code a code consisting of dots and dashes that code for each letter of the alphabet

N

NAND gate a combination of an AND gate followed by a NOT gate

National Grid network that carries electricity from power stations across the country (it uses cables, transformers and pylons)

near-Earth object asteroid, comet or large meteoroid whose orbit crosses the Earth's orbit

neutral (wire) provides a return path for the current in a mains supply to a local electricity substation

neutrons small particle which does not have a charge found in the nucleus of an atom

newtons unit of force (abbreviated to N)

noise unwanted signals

NOR gate a combination of an OR gate followed by a NOT gate

NOT gate a logic gate whose output is opposite to its input

nuclear equation equation showing changes to the nuclei in a nuclear reaction

nuclear power stations power stations using the energy produced by nuclear fission to generate heat

nucleons protons and neutrons (both found in the nucleus)

nucleus central part of an atom that contains protons and neutrons

O

ohm the unit used to measure electrical resistance

optical fibres very thin glass fibres that light travels along by total internal reflection

OR gate a logic gate which delivers an output if any input terminal is on

orbit the path taken by a planet around a sun, a moon around a planet or a satellite around another body in space

oscilloscope a device that displays a line on a screen showing regular changes (oscillations) in something. An oscilloscope is often used to look at sound waves collected by a microphone

ozone layer layer of the Earth's atmosphere that protects us from ultraviolet rays

P

paddle shift controls controls attached to the steering wheel of a car so that the driver can use them without taking their eyes off the road

paddles charged plates in a defibrillator that are placed on the patient's chest

parabolic shaped like a parabola, which looks a bit like an opened umbrella

parallelogram of forces a method of finding the resultant of two forces (or other vectors)

petrol volatile mixture of hydrocarbons used as a fuel

(in) phase when two waves are 'in step' with each other; crests coincide and troughs coincide

photocell a device which converts light into electricity

pitch whether a sound is high or low on a musical scale

plutonium a radioactive metal often formed as a bi-product from a nuclear power station – sometimes used as a nuclear fuel

p–n junction the boundary between two special types of silicon in a photocell and other electronic components

polar orbit a satellite orbit that passes over Earth's North and South poles

polarised (light) light in which the oscillations are confined to one plane only

Polaroid a material that absorbs light except that polarised in one particular plane, producing polarised light

pollution contaminating or destroying the environment as a result of human activities

potential difference another word for voltage (a measure of the energy carried by the electric current)

potential divider a combination of two resistors which allows an output voltage that is a fraction of the input voltage

power the rate that a system transfers energy, power is usually measured in watts (W)

pressure the force acting normally per unit area of a surface. Pressure = $\frac{force}{area}$. Unit: pascals (Pa) or N/m²

pressure wave vibrating particles in a longitudinal wave creating pressure variations

primary safety features help to prevent a crash, e.g. ABS brakes, traction control

primary coil the input coil of a transformer

principal axis the axis, perpendicular to the face of a lens, that passes through the optical centre

probe unmanned space vehicle designed to travel beyond the Earth's orbit

projectile any object thrown or fired in the Earth's gravitational field

projector an optical instrument that produces an enlarged image on a screen

protons small positive particles found in the nucleus of an atom

P wave longitudinal seismic wave capable of travelling through solid and liquid parts of the Earth

R

radiation thermal energy transfer which occurs when something hotter than its surroundings radiates heat from its surface

radio waves non-ionising waves used to broadcast radio and TV programmes

radioactive material which gives out radiation

radioactive waste waste produced by radioactive materials used at nuclear power stations, research centres and some hospitals

radiocarbon dating method of dating some old artefacts using carbon-14

radioisotope isotope of an element that is radioactive

radiotherapy using ionising radiation to kill cancer cells in the body

random having no regular pattern

range the horizontal distance covered by an object after it has been thrown or fired

rarefactions particles further apart than usual, decreasing pressure

reaction force when an object feels a force it pushes back with an equal reaction force in an opposite direction

reaction time the time it takes for a driver to step on the brake after seeing an obstacle

real (image) an image that can be projected onto a screen; light actually passes through it

receiver device which receives waves, e.g. a mobile phone

recharging battery being charged with a flow of electric current

recoils bounces or pushes back

red shift when lines in a spectrum are redder than expected – if an object has a red shift it is moving away from us

refraction a change in speed, and usually direction, when light passes from one medium to another, e.g. from air to glass or water

refractive index the ratio of the speed of light in a vacuum (or air) to the speed of light in a medium

relative speed the speed of one moving object with respect to another

relay an electronic switch which allows a very small current in one circuit to switch a larger current in another circuit

renewable energy energy that can be replenished at the same rate that it's used up e.g. biofuels

repel move away, for example, like charges repel

resistance measurement of how hard it is for an electric current to flow through a material

resistor a conductor that reduces the flow of electric current

resultant force the combined effect of forces acting on an object

rheostat a variable resistor

rocket vehicle that travels into space carrying its own oxygen supply

S

satellite a body orbiting a larger body, e.g. communications satellites orbit the Earth

scalar a quantity having magnitude but no direction

scattered moved in random directions

seat belt harness worn by occupants of motor vehicles to prevent them from being thrown about in a collision

secondary coil the output coil of a transformer

sensor device that detects a change in the environment

shock occurs when a person comes into contact with an electrical energy source so that electrical energy flows through a portion of the body

shutter in a camera, it opens and closes very quickly to let light into the camera

slip rings allow the transfer of current from a rotating coil in an AC generator

smoke detector device to detect smoke, some forms of which contain a source of alpha radiation

Solar System the collection of planets and other objects orbiting around the Sun

solar power energy provided by the Sun

Glossary

sparks type of electrostatic discharge briefly producing light and sound

specific heat capacity the amount of energy needed to raise the temperature of 1 kg of a substance by 1 degree Celsius

specific latent heat the amount of energy needed to change the state of a substance without changing its temperature; for example the energy needed to change ice at 0 °C to water at the same temperature

speed how fast an object travels: speed = distance ÷ time

speed–time graph a plot of how the speed of an object varies with time

star bright object in the sky which is lit by energy from nuclear reactions

stopping distance sum of the thinking and braking distances

straight line line of constant gradient

stratosphere a layer in the atmosphere starting at 15 km above sea level and extending to 50 km above sea level; the ozone layer is found in the stratosphere

S wave transverse seismic wave capable of travelling through solid but not liquid parts of the Earth

T

temperature a measure of the degree of hotness of a body on an arbitrary scale

terminal speed or velocity the top speed reached when drag matches the driving force

therapy treatment of a medical problem

thermistor an electronic device whose resistance changes with temperature

thermogram a picture showing differences in surface temperature of a body

thinking distance distance travelled while the driver reacts before braking

threshold voltage the voltage needed to switch on a circuit

total internal reflection complete reflection of a light ray within glass when the ray hits the glass/air boundary at an angle which is greater than the critical angle

trajectory the path of a projectile

transformers devices which can change the voltage and current of electricity

transistor an electronic component used to amplify or switch electronic signals

transmitter device which transmits waves, e.g. a mobile-phone mast

transverse (wave) wave in which the vibrations are at right angles to the direction in which the wave travels

tread pattern on part of tyre that comes in contact with road surface to provide traction

troughs lowest points of a wave

truth table a mathematical way of describing the behaviour of logic gates

tsunami huge waves caused by earthquakes – can be very destructive

tumour abnormal mass of tissue that is often cancerous

turbine device for generating electricity – the turbine moves through a magnetic field and electricity is generated

U

ultrasound high-pitched sounds which are too high for detection by human ears

ultraviolet (UV) radiation electromagnetic waves given out by the Sun which damage human skin

unbalanced (forces) forces acting in opposite directions that are unequal in size

universe the whole of space

unstable (nucleus) liable to decay

uranium radioactive element with a very long half-life used in nuclear power stations

V

vacuum space containing hardly any particles

variable resistor a device whose resistance can change, often used in volume controls and dimmer switches

velocity how fast an object is travelling in a certain direction: velocity = displacement ÷ time

virtual image image formed on the same side of the lens as the object; a virtual image formed by reflection can be seen but cannot be projected onto a screen

volt unit used to measure voltage

voltage the potential difference across a component or circuit

voltmeter instrument used to measure voltage or potential difference

W

watt (W) a unit of power, 1 watt equals 1 joule of energy being transferred per second

wave oscillatory motion

wavelength distance between two wave peaks

weight force on an object due to gravitational attraction (= mg). Unit: N

work done the product of the force and distance moved in the direction of the force

X

x-rays ionising electromagnetic waves used in x-ray photography (where x-rays are used to generate pictures of bones)

Collins Revision

New GCSE Physics

Exam Practice Workbook

Higher

For OCR Gateway B

Exam tips

The key to successful revision is finding the method that suits you best. There is no right or wrong way to do it.

Before you begin, it is important to plan your revision carefully. If you have allocated enough time in advance, you can walk into the exam with confidence, knowing that you are fully prepared.

Start well before the date of the exam, not the day before!

It is worth preparing a revision timetable and trying to stick to it. Use it during the lead up to the exams and between each exam. Make sure you plan some time off too.

Different people revise in different ways and you will soon discover what works best for you.

Remember

There is a difference between *learning* and *revising*.

When you revise, you are looking again at something you have already learned. Revising is a process that helps you to remember this information more clearly.

Learning is about finding out and understanding new information.

Using the Workbook

This Workbook allows you to work at your own pace and check your answers using the detachable Answer section on pages 137–148. In addition to the exam practice questions, the Workbook also contains questions that require longer answers (Extended response questions). You will find one question that is similar to these in each section of your written exam papers. The model answers supplied for these questions give guidance about the content that should be included, but do not necessarily provide a complete response for the questions concerned.

Some general points to think about when revising

- Find a quiet and comfortable space at home where you won't be disturbed. You will find you achieve more if the room is ventilated and has plenty of light.

- Take regular breaks. Some evidence suggests that revision is most effective when tackled in 30 to 40 minute slots. If you get bogged down at any point, take a break and go back to it later when you are feeling fresh. Try not to revise when you're feeling tired. If you do feel tired, take a break.

- Use your school notes, textbook and this Revision guide.

- Spend some time working through past papers to familiarise yourself with the exam format.

- Produce your own **summaries** of each module and then look at the summaries in this Revision guide at the end of each module.

- Draw mind maps covering the key information on each topic or module.

- Review the **Grade booster checklists** on pages 128–133.

- Set up revision cards containing condensed versions of your notes.

- Prioritise your revision of topics. You may want to leave more time to revise the topics you find most difficult.

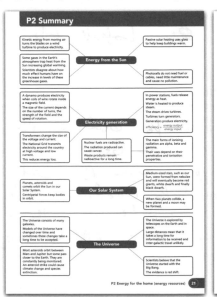

Heating houses

1 Kelly opens the front door on a very cold morning. Her mother complains that the house is getting cold. Use your ideas about energy flow to explain why the house gets cold.

As heat travels from a hot area to a colder area, the heat inside the house will transfer to the cold air outside, causing the house to cool down **[2 marks]**

2 Explain the difference between temperature and heat.

Temperature is a ~~strong~~ measurement of hotness on an arbitrary scale while heat is measured on an absolute scale **[2 marks]**

3 A police helicopter uses a thermal imaging camera to take a picture at night. It is looking for a car that has recently been abandoned in a field after a high speed chase. Describe how the thermogram can help locate the car.

The heat from the engine will still be visible when contrasted with the rest of the cold field and so it will be easily identifiable on the thermogram **[3 marks]**

4 a Finish the sentence.

The energy needed to raise the temperature of 1 kg of a material by 1 °C is known as the

Specific ~~heat~~ Heat Capacity **[1 mark]**

b i Jamie heats 500 g of seawater in a beaker from 20 °C to 90 °C. The specific heat capacity of seawater is 3900 J/kg °C. How much energy is needed to heat the sea water?

3900 J/kg °C × 0.5 kg = 1,950 J/°C
1,950 J/°C ÷ 70 °C = 27.86 J (2 s.f.) **[2 marks]**

ii The energy supplied is greater than this. Suggest where this additional energy has been used.

Heating the container and the environment **[1 mark]**

5 a What physical quantity is measured in units of J/kg?

Specific ~~heat~~ ~~cap~~ Latent Heat **[1 mark]**

b When iron changes from solid to liquid, energy is transferred but there is no temperature change until all of the solid has changed into liquid. Use your ideas about molecular structure to explain why there is no change in temperature at iron's melting point.

Energy is needed to weaken the intermolecular forces to make the iron become liquid. This energy is supplied via the heat and so all of it will be used to weaken these ~~forces~~ bonds and none of it will be used to raise the temperature until the iron is completely liquid. **[2 marks]**

74 P1 Energy for the home

Keeping homes warm

1 a 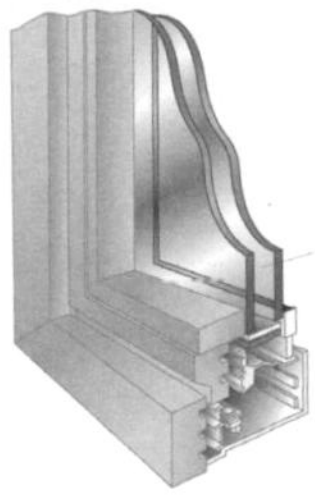 The diagram shows a section through a double glazed window. Michael says that it is just as effective to use a piece of glass twice the thickness. Use your ideas about energy transfer to explain why double glazing is better.

....

....

.... [2 marks]

b New homes are built with insulation blocks in the cavity between the inner and outer walls. The blocks have shiny foil on both sides.

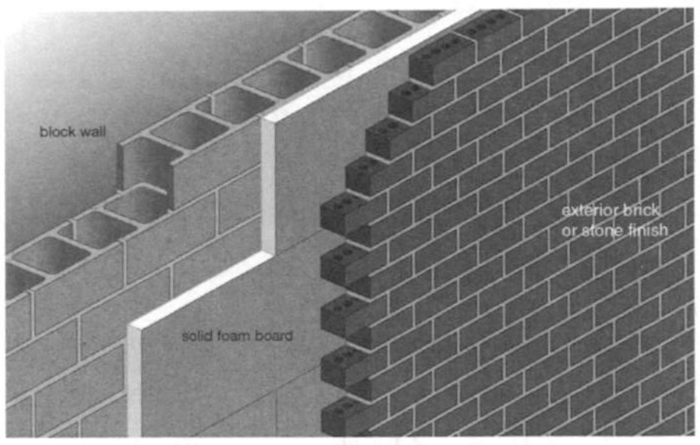

block wall

exterior brick or stone finish

solid foam board

D–C

i Explain how the insulation blocks reduce energy transfer by conduction and convection.

....

....

....

.... [3 marks]

ii Explain how the shiny foil helps to keep a home warmer in winter and cooler in summer.

....

....

....

2 a A brick in a wall is a better conductor of heat than air in the cavity between the walls. Explain why.

....

....

.... [2 marks]

B–A*

b Hot air rises. Explain why.

....

....

.... [2 marks]

3 Dan heats his house with coal fires. He is told that his fires are 32% efficient.

a Explain what is meant by 32% efficient.

....

.... [1 mark]

b Dan pays £9.50 for a 25 kg bag of coal. How much of that money is usefully used in heating his house?

D–C

....

....

.... [2 marks]

c Suggest why coal fires are so inefficient?

....

.... [1 mark]

A spectrum of waves

1 The diagram shows a transverse wave.

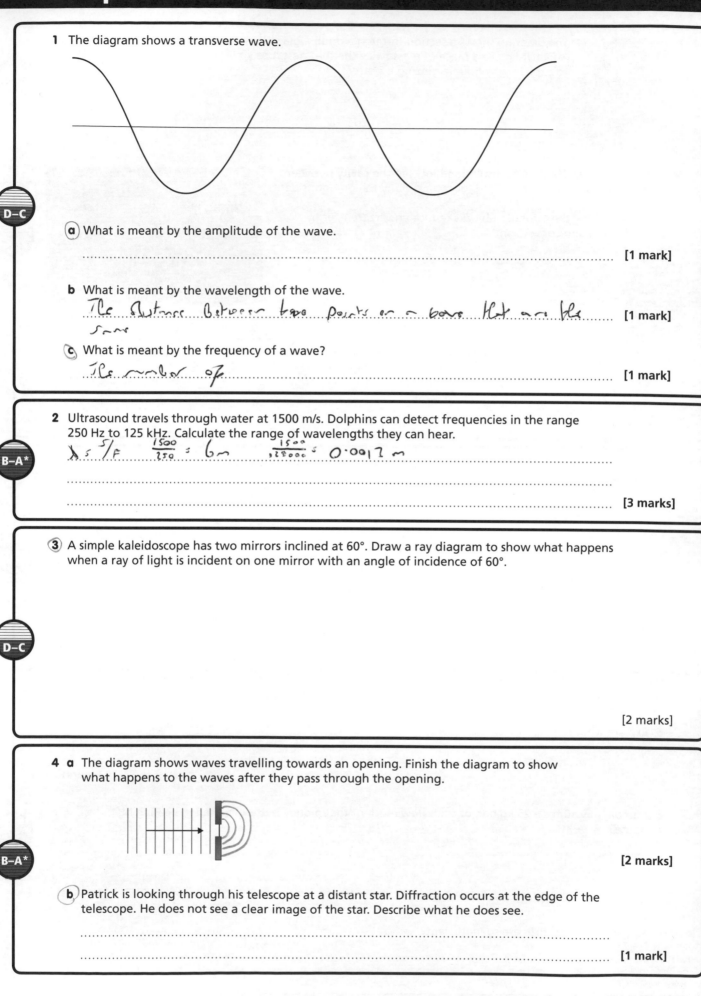

a What is meant by the amplitude of the wave.

.... **[1 mark]**

b What is meant by the wavelength of the wave.

The distance between two points on a wave that are the same **[1 mark]**

c What is meant by the frequency of a wave?

The number of .... **[1 mark]**

2 Ultrasound travels through water at 1500 m/s. Dolphins can detect frequencies in the range 250 Hz to 125 kHz. Calculate the range of wavelengths they can hear.

$\lambda = \dfrac{v}{f}$ $\dfrac{1500}{250} = 6m$ $\dfrac{1500}{125000} = 0.0012\,m$

....

.... **[3 marks]**

3 A simple kaleidoscope has two mirrors inclined at 60°. Draw a ray diagram to show what happens when a ray of light is incident on one mirror with an angle of incidence of 60°.

[2 marks]

4 a The diagram shows waves travelling towards an opening. Finish the diagram to show what happens to the waves after they pass through the opening.

[2 marks]

b Patrick is looking through his telescope at a distant star. Diffraction occurs at the edge of the telescope. He does not see a clear image of the star. Describe what he does see.

....

.... **[1 mark]**

Light and lasers

1 Why is Morse code an example of a digital signal and not an analogue signal?

....

....

.... [2 marks]

D–C

2 Tina's science teacher shines a laser light onto a screen. It is brighter than the white light from the laboratory lights. Use your ideas about frequency to explain the difference between laser light and white light. You may draw diagrams to help you answer the question.

B–A*

....

....

.... [3 marks]

3 a The diagrams show three rays of light travelling from water into air. The three angles of incidence are (x) smaller than the critical angle (y) equal to the critical angle (z) larger than the critical angle.

D–C

i Finish the diagrams to show what happens to the rays of light after they meet the water/air boundary. [4 marks]

ii Show clearly, on the correct diagram, the critical angle. Label it c. [1 mark]

b Sonia's doctor wants to look at the inside of her stomach. The doctor does so without using surgery. Describe how the doctor can see inside Sonia's stomach.

....

....

....

....

....

....

....

....

.... [3 marks]

B–A*

Cooking and communicating using waves

1 a Microwave ovens take less time to cook food than normal ovens. Suggest why.

....
.... **[1 mark]**

b Microwaves are suitable to communicate with space craft thousands of kilometres away, but mobile phones often cannot receive a signal just a few kilometres from the nearest transmitter. Why do microwave signals seem to work better in space than they do on Earth?

....
....
....
.... **[2 marks]**

D–C

2 The diagram shows the electromagnetic spectrum.

| radio | microwave | infrared | visible | ultraviolet | X-ray | gamma ray |

a Which part of the electromagnetic spectrum transfers the most energy?

.... **[1 mark]**

B–A*

b An electric iron and an element from an electric fire both emit infrared radiation. The iron transfers less energy than the element. Explain how the wavelength of radiation from the iron differs from the wavelength of radiation from the element.

....
....
....
....
.... **[2 marks]**

3 The diagram shows a transmitter on top of a hill. It transmits microwave mobile phone signals as well as radio and television signals.

B–A*

The house, behind the other hill, can receive both radio and television signals but there is no mobile phone reception. Explain why.

....
....
....
....
.... **[2 marks]**

Data transmission

1 a The change from analogue to digital transmission of television signals began in the United Kingdom in 2009. Write down one advantage of digital television.

.... **[1 mark]**

D–C

2 A ray of laser light is shone into one end of an optical fibre.

Finish the path of the ray as it passes into, through and out of the optical fibre. **[2 marks]**

D–C

3 A multinational company has thousands of computers which transmit data continually. The data is not transmitted using analogue signals. Explain the advantages of transmitting data digitally. You may use diagrams to illustrate your answer

B–A*

....

....

....

....

....

....

....

....

....

....

....

.... **[5 marks]**

Wireless signals

1 a Radio waves are refracted in the upper atmosphere. What happens to the amount of refraction if the frequency of the radio wave is decreased?

....

....

....

....

....

.... **[1 mark]**

b Why does the microwave beam sent by a transmitting aerial towards a satellite in orbit have to be **focused**?

....

....

....

....

....

.... **[1 mark]**

2 When Jenny is watching her television, she notices that there is a faint second picture slightly offset to the main picture.

Finish the sentence to explain why there is this 'ghost' picture.

Choose words from this list.

absorbed   **dispersed**   **reflected**   **refracted**

The aerial has received a direct signal from the transmitter and a signal that has been

.... **[1 mark]**

3 a Jenny listens to her favourite radio station. Every so often, she notices that she can hear a foreign radio station as well. Put ticks (✓) in the **two** boxes next to the statements that explain why this happens. **[1 mark]**

The foreign radio station is broadcasting on the same frequency. ☐

The foreign radio station is broadcasting with a more powerful transmitter. ☐

The radio waves travel further because of weather conditions. ☐

Jenny's radio needs new batteries. ☐

b Explain the benefits Jenny will get if she replaces her radio with a DAB radio?

....

....

....

....

....

.... **[1 mark]**

D-C

B-A*

D-C

D-C

B-A*

Stable Earth

1 a P waves and S waves are two of the waves which travel through the Earth after an earthquake. Finish the table by putting a tick (✓) in the correct box or boxes next to the description of the wave. The first one has been done for you.

[3 marks]

description	P wave	S wave
pressure wave	✓	
transverse wave		
longitudinal wave		
travels through solid		
travels through liquid		

D–C

b How do scientists use the properties of P waves and S waves to find out the **size** of the Earth's core?

....
....
.... **[2 marks]**

B–A*

c How do scientists use the properties of P waves and S waves to find out the **structure** of the Earth's core?

....
....
.... **[2 marks]**

2 Sandy wants to sunbathe and get a good tan.

a She is told that if she goes out in the Sun without sunscreen, she will burn in 15 minutes. How long can she safely sunbathe for if she uses a sunscreen with SPF 20?

....
.... **[2 marks]**

D–C

b The Earth's atmosphere contains a layer of ozone. This layer protects us from the effects of the Sun. When scientists first started measuring the thickness of the ozone layer, their results were unexpected. The layer was thinner than they thought it should be. They replaced all of their instruments. What should scientists do to confirm their results?

....
....
.... **[1 mark]**

3 Dave reads a newspaper article that claims CFCs are not responsible for the depletion of the ozone layer. He tells his friends at school that the science teachers have not been teaching them correctly about ozone depletion. Explain why Dave should not have believed what he read in the newspaper.

B–A*

....
....
....
.... **[2 marks]**

Many people now use mobile phones.

Jenny is walking in the hills of North Wales and her brother is walking around the Lake District. They keep in touch using their mobile phones. Explain reasons why the signal strength on both their phones changes, even as they walk only a short distance. Why do some people think it would be better if Jenny were to text her brother instead?

❶ The quality of written communication will be assessed in your answer to this question.

....
....
....
....
....
....
....
....
....
....
....
....
....
....
....
....
....
....
....
....
....
....
....
....
....
....
....
....
....
....
....
.... [6 marks]

Collecting energy from the Sun

1 a Write down **four** advantages of using photocells.

....

....

....

.... [4 marks]

D–C

b A p-n junction is made from two pieces of silicon. Explain how the two pieces of silicon are different and what causes the difference.

....

....

.... [3 marks]

B–A*

2 Name two methods of increasing the output from a photocell.

....

.... [2 marks]

3 a This is a question about passive solar heating.

During the day, short wavelength radiation from the Sun passes through the glass in the large window and warms the room.

Explain how the glass keeps the room warm during the night

.... [1 mark]

B–A*

b The diagram represents the electromagnetic spectrum.

X-rays	ultraviolet	visible light	infrared	radio

i Write the letter S on the diagram to show the wavelength of radiation from the Sun that is absorbed by plants in a greenhouse.   [1 mark]

ii Write the letter P on the diagram to show the wavelength of radiation that is re-radiated from the plants in a greenhouse.

.... [1 mark]

4 a A wind turbine transfers the kinetic energy of the wind into electricity. What does the amount of electricity produced depend on?

.... [1 mark]

b Write down two disadvantages and two advantages of generating electricity using wind turbines.

Advantages

....

....

Disadvantages

....

.... [4 marks]

D–C

Generating electricity

1 **a** The diagram shows a model dynamo. When the coil is spun, a current is produced.

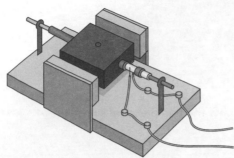

Write down two ways in which the size of the current can be increased.

....

.... **[2 marks]**

b A model generator consists of a coil of wire rotating between the poles of a magnet. How is the structure of a generator at a power station different from the model generator?

....

....

.... **[1 mark]**

c The oscilloscope trace shows the output from a generator. Each division of the timescale is 5 milliseconds.

i Mark on the diagram the period of the alternating voltage. **[1 mark]**

ii What is the peak voltage?

.... **[1 mark]**

2 A power station is about 30% efficient. If the fuel provides 200 MJ of energy, how much electrical energy will be produced?

....

....

....

....

....

....

.... **[1 mark]**

Global warming

1 a How do the greenhouse gases cause the temperature of the Earth to increase?

.... [1 mark]

b Which of the greenhouse gases is the most significant ?

.... [1 mark]

2 a How has deforestation affected the levels of carbon dioxide in the atmosphere?

.... [1 mark]

b Write down four natural sources of carbon dioxide.

.... [4 marks]

c Give three sources of methane gas caused by human activity.

.... [3 marks]

3 a Describe how electromagnetic radiation from the Sun can be trapped by gases in the Earth's atmosphere.

....
.... [3 marks]

b Dust in the atmosphere can cause either an increase or a decrease in the Earth's temperature. Explain how this is possible.

Increase in temperature is caused by:

.... [1 mark]

Decrease in temperature is caused by:

.... [1 mark]

4 a Most scientists agree that the average temperature of the Earth is increasing but what do they disagree on?

.... [1 mark]

b On what basis should governments make decisions on what action to take on global warming?

.... [1 mark]

c Describe two consequences of global warming.

....
.... [2 marks]

5 Say whether the following statements are based on scientific fact or opinion.

Average temperatures have risen by 0.8 degree Celsius around the world since 1880 according to NASA.

.... [1 mark]

The recent extreme weather, such as heat waves and tropical storms, is caused by climate change say some experts.

.... [1 mark]

As illustrated in Ice Core data from the Soviet Station Vostok in Antarctica, CO_2 concentrations in Earth's atmosphere has been rising for 18 000 years.

.... [1 mark]

Fuels for power

1 a Each of the headlamp bulbs in Sammy's car is connected to a 12 V battery. When she switches on the headlamps, a current of 2 A passes through the bulb. Calculate the power rating of the bulb.

....

....

....

.... **[2 marks]**

b In her home, Sammy uses a 2.5 kW kettle for $\frac{1}{2}$ hour each day. Electricity costs 12p per kWh. How much does it cost Sammy each day to use her kettle?

....

....

.... **[3 marks]**

c Sammy sets her dishwasher to work overnight. Why is electricity cheaper during the night?

....

....

.... **[1 mark]**

d Sammy knows she needs to keep her electricity use down below 2 kWh per night. She has a 4 KW dryer, how long can she use it for?

....

.... **[1 mark]**

2 a List three factors which should be taken into consideration when deciding on which energy source to use in a particular situation.

....

....

.... **[3 marks]**

b Some people believe that the UK needs more nuclear power stations. Others believe that renewable fuel can provide all of our energy requirements. What information could scientists and engineers provide to help the government decide?

....

....

.... **[3 marks]**

3 The National Grid distributes electricity around the country at 400 000 V.

a Give **two** reasons why such high voltages are used.

.... **[2 marks]**

b How does using such a high voltage affect the current in the wires?

.... **[1 mark]**

c How would this help with energy losses?

.... **[1 mark]**

Nuclear radiations

1 a Answer true or false to each of the following statements about alpha, beta and gamma radiation.

Gamma radiation is stopped by paper.	
Alpha radiation has a range of a few centimetres in air.	
Beta radiation comes from the nucleus of an atom.	
Beta radiation can be absorbed by a thin sheet of paper.	

[4 marks]   D–C

b i Atoms are neutral because they contain the same number of positive protons and negative electrons. Explain how negative and positive ions are formed when an atom is exposed to radiation.

negative ions are formed by:

.... [1 mark]

positive ions are formed by:

.... [1 mark]

ii Describe the effects ionisation can have on the cells of the human body.

....

.... [2 marks]   B–A*

2 Gamma radiation is used to sterilise medical instruments. It has other medical uses as well.

a Write down one other medical use for gamma radiation.

.... [1 mark]

b What property of gamma radiation makes it suitable?

.... [1 mark]   D–C

3 A source of alpha radiation is used in a smoke alarm. Explain how a smoke alarm works.

....

....

.... [3 marks]

4 Radioactive waste must be stored securely for possibly thousands of years.

a Why must it be stored for so long?

....

.... [1 mark]   D–C

b Some people are worried that terrorists may make a nuclear bomb from nuclear waste. Discuss whether or not we should have any concerns about terrorists obtaining nuclear waste.

....

....

.... [3 marks]   B–A*

Exploring our Solar System

1 a There are many objects in our Solar System. Describe the following objects:

i a star

.... [1 mark]

ii a planet

.... [1 mark]

iii a meteor

.... [1 mark]

b Our Solar System is part of a galaxy and scientists think there may be a black hole at the centre of our galaxy.

i What is a galaxy?

.... [1 mark]

ii Describe a black hole.

.... [1 mark]

c The diagram represents the Moon in orbit around Earth.

i What is the name of the force that keeps the Moon in orbit around Earth?

.... [1 mark]

ii Add an arrow to the diagram to show the direction in which this force acts on the Moon.

[1 mark]

2 a When astronauts work outside a spacecraft, they have to wear special helmets with Sun visors. Why do the helmets need special visors?

.... [1 mark]

b NASA is planning to send a manned spacecraft to another planet after 2020. It is expected to cost £400 billion.

Which planet will the spacecraft go to? Explain the reason for your answer.

Planet:

.... [1 mark]

Explanation:

.... [1 mark]

3 Proxima Centauri is 4.22 light-years away from us.

Explain what is meant by the term light-year.

.... [1 mark]

P2 Living for the future (energy resources)

Threats to Earth

1 a Most asteroids orbit the Sun in a belt between two planets. Which two planets?

.... [2 marks] D–C

b Explain why asteroids have not joined together to form another planet.

....

.... [2 marks] B–A*

c Describe two pieces of evidence scientists have discovered that support the theory that asteroids have collided with Earth in the past.

....

.... [2 marks] D–C

2 There is evidence to suggest our Moon was a result of the collision between two planets. The iron core of the other planet melted and joined with the Earth's core.

Describe two pieces of evidence to support this theory. B–A*

....

.... [2 marks]

3 The diagram shows the orbits of two bodies orbiting the Sun. One is a planet, the other is a comet.

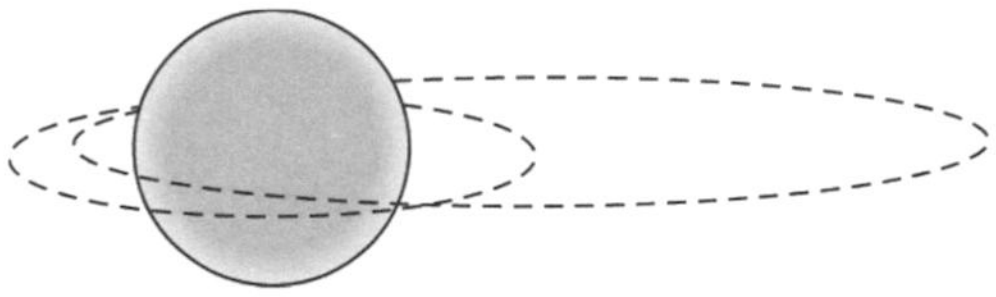

D–C

a label the comet's orbit with the letter C. [1 mark]

b Write the letter X to show where on the orbit the comet is travelling at its fastest. [1 mark] B–A*

c Why does a comet's tail always point away from the Sun?

.... [1 mark] D–C

4 Scientists are constantly updating information on the paths of near-Earth objects (NEO)s.

a Why is it important to constantly monitor the paths of NEOs? D–C

....

.... [2 marks]

b i Scientists may want to change the course of a NEO. Explain how this may be done.

....

.... [2 marks]

ii what would they need to take into consideration when planning to change the course of a NEO? B–A*

....

.... [2 marks]

The Big Bang

1 a The Universe is expanding. Galaxies in the Universe are moving at different speeds. Which galaxies are moving the fastest?

.... [1 mark]

b Scientists know how fast galaxies are moving because they measure red shift.

Explain what is meant by red shift.

....

....

.... [3 marks]

c How does red shift provide information about the speed of a galaxy?

....

.... [1 mark]

2 A star starts its life as a swirling cloud of gas and dust.

a Describe what happens to this cloud to produce a glowing star.

....

....

....

.... [4 marks]

The end of a star's life depends on how big it is.

b What happens to a medium sized star, like our Sun, at the end of its life?

....

....

....

.... [4 marks]

3 Models of our universe have changed a lot over time.

a How did Galileo contribute to the changes?

....

.... [1 mark]

b Why did it take a long time for Galileo's ideas to be accepted?

....

.... [1 mark]

c What did Isaac Newton contribute to the model we have today?

....

.... [1 mark]

P2 Living for the future (energy resources)

P2 Extended response question

Scientists have been making observations of our universe for a very long time. They have gathered a lot of information on the life cycle of the stars they observe.

Describe, in detail, the life cycle of a medium-sized star such as our Sun. Explain why scientists describe our Sun as a second generation star.

❶ The quality of written communication will be assessed in your answer to this question.

....
....
....
....
....
....
....
....
....
....
....
....
....
....
....
....
....
....
....
....
....
....
....
....
....
....
....
....
....
....
....
....
....

[6 marks]

Speed

1 The graph shows Ashna's walk to her local shop and home again.

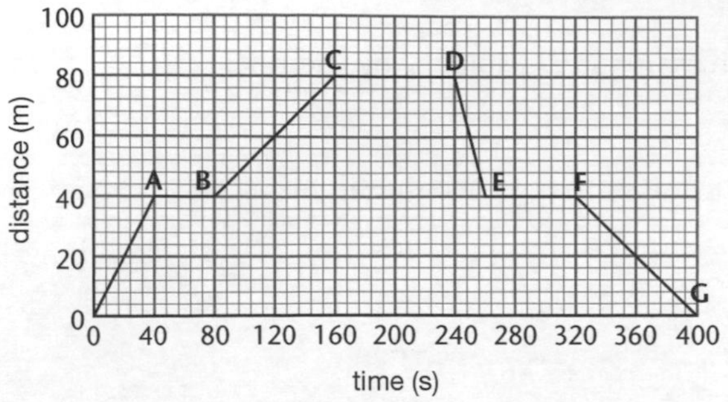

D–C

a How long did she spend in the shop?

.... [1 mark]

b How far is the shop from her home?

.... [1 mark]

c During which part of her journey did she walk fastest?

.... [1 mark]

d Calculate Ashna's speed between 0 and A.

....

....

.... [2 marks]

B–A*

e Calculate Ashna's speed between F and G.

....

....

.... [2 marks]

2 A cycle track is 500 m long. Imran completes 10 laps. He cycles at an average speed of 45 km/h.

a How long did he take to complete ten laps?

....

....

.... [3 marks]

b Imran put on a spurt in the last lap, completing it in 35 s. What was his average speed, in m/s, for the last lap?

....

....

.... [2 marks]

B–A*

c Calculate Imran's average speed for the first nine laps.

....

....

.... [3 marks]

Changing speed

1 Darren is riding his bicycle along a road. The speed–time graph shows how his speed changed during the first minute of his journey.

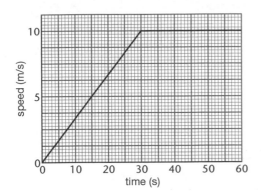

a Darren makes the same journey the next day but:
- increases his speed at a steady rate for the first 20 s, reaching a speed of 10 m/s
- travels at a constant speed for 10 s
- slows down at a steady rate for 15 s to a speed of 5 m/s
- travels at a constant speed of 5 m/s.

Plot the graph of this journey on the same axes. **[4 marks]**

b How could you calculate the distance Darren travelled in the first minute of his original journey?

.... **[1 mark]**

D–C

2 The way in which the speed of a car changes over a 60 s period is shown in the table.

a Plot a speed–time graph for the car using the axes given.

time in s	speed in m/s
0	0
5	5
10	10
15	15
20	15
25	15
30	15
35	15
40	15
45	11
50	7
55	3
60	0

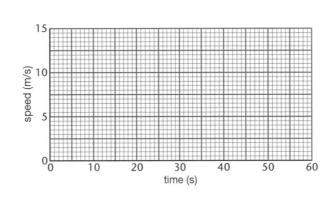

[4 marks]

b Calculate the deceleration between 40 and 55 seconds.

....
....
....
.... **[3 marks]**

D–C

c Use the graph to calculate the distance travelled by the car.

....
.... **[2 marks]**

B–A*

3 A cheetah accelerates at 6 m/s². How fast is it moving after 5 seconds, starting from rest?

....
.... **[2 marks]**

B–A*

Forces and motion

1 A car of mass 500 kg accelerates steadily from 0 to 40 m/s in 20 s.

D–C

    a What is the size of the resultant force which produces this acceleration?

....

.... [2 marks]

    b Calculate the acceleration of the car if the resultant force is increased to 1250 N.

....

.... [2 marks]

2 Helen is driving her car on a busy road when the car in front brakes suddenly. She puts her foot firmly on her brake pedal and just manages to stop without hitting the car in front.

    a Write down two things, apart from speed, that could increase a driver's thinking distance.

....

.... [2 marks]

D–C

    b Later that day Helen is driving at high speed on a motorway.

      i How will Helen's thinking distance change?

.... [1 mark]

      ii Explain why.

....

.... [2 marks]

    c Explain the effect worn tyres have on braking distance.

B–A*

.... [2 marks]

3 Finish the graphs to show how thinking distance and braking distance change with speed. Some points have been plotted for you.

[3 marks]

B–A*

P3 Forces for transport

Work and power

1 Hilary lifts a parcel of weight 80 N onto a shelf 2 m above the ground.

   a Calculate the amount of work Hilary does.

   ....

   .... [2 marks]

   b i How high could Hilary lift a parcel of weight 60 N if she did the same amount of work in 1.5 s?

   ....

   .... [2 marks]

   ii What is Hilary's power output now?

   ....

   .... [2 marks]

D–C

2 Chris and Abi both have a mass of 60 kg. They both run up a flight of stairs 3 m high. Chris takes
  8 s and Abi takes 12 s.

   a What can you say about the power of Chris and Abi?

   .... [1 mark]

   b Calculate Chris' weight. (Take g = 10 N/kg.)

   ....

   .... [2 marks]

D–C

   c Calculate Chris' power.

   ....

   .... [2 marks]

   d Calculate Abi's power.

   ....

   .... [2 marks]

B–A*

   e Chris tells Abi that on Jupiter they would each weigh 1560 N. What is the gravitational field
     strength on Jupiter?

   .... [1 mark]

3 The table shows the fuel consumption of three cars in miles per gallon (mpg).

Car fuel	Consumption (mpg)
A	48
B	34
C	28

   a Which car has the best fuel consumption?

   .... [1 mark]

   b Which car is likely to be most powerful?

   .... [1 mark]

D–C

   c We should keep our fuel consumption to a minimum to protect the environment. Why?

   ....

   ....

   .... [3 marks]

Energy on the move

1 Use the data about fuel consumption to answer the following questions.

Car	Fuel	Engine size in litres	Miles per gallon	
			urban	non-urban
Renault Megane	petrol	2.0	25	32
Land Rover	petrol	4.2	14	24

D–C

a Suggest why fuel consumption is better in non-urban conditions.

....

.... [2 marks]

b How many gallons of petrol would a Land Rover use on a non-urban journey of 96 miles?

.... [2 marks]

c Suggest why a Renault Megane uses less fuel for the same journey?

.... [1 mark]

B–A*

d Explain how driving style can affect fuel consumption.

....

....

.... [3 marks]

2 a How do battery-powered cars pollute the environment?

.... [1 mark]

D–C

b Give one advantage and one disadvantage of solar-powered cars compared to battery-powered cars.

....

.... [2 marks]

3 Sam is driving a car of mass 1200 kg at a speed of 20 m/s.

a Calculate the kinetic energy of the car.

....

....

.... [2 marks]

B–A*

b When Sam suddenly applies the brakes, the car travels 32 m before it stops. Sam suggests that the braking distance would be 16 m if he was travelling at 10 m/s. Explain why Sam is not correct.

....

.... [2 marks]

Crumple zones

1 Kevin was involved in an accident on the M1 motorway. Luckily he was not seriously hurt but his car was badly damaged. Kevin's car had crumple zones, seat belts, airbags and ABS brakes.

    a Finish the table by describing how each feature helps to reduce Kevin's injuries.

Safety feature	How it works
seat belt	
crumple zones	
air bag	

[3 marks]

D–C

    b Describe how ABS brakes work.

....

....

.... [2 marks]

B–A*

2 Marie was a passenger in a car travelling at 25 m/s when the driver braked sharply to avoid a dog that had run into the road. Marie was wearing her seat belt and this brought her to a stop in 0.5 s. Marie's mass is 55 kg.

    a Calculate the average force the seat belt exerted on her body.

....

....

.... [2 marks]

D–C

    b If Marie had not been wearing her seat belt, she would have hit the windscreen which would have brought her head to a stop in 0.002 s. Calculate what the average force on her head would have been.

....

....

.... [2 marks]

3 To minimise injury the forces acting on the people in a car during a car accident must be as small as possible.

    a Explain why this means safety features must reduce the deceleration of the car on impact.

....

.... [2 marks]

B–A*

    b Explain how one safety feature is designed to reduce the deceleration of the car on impact.

....

.... [2 marks]

Falling safely

1 Charlie drops a golf ball and a ping pong ball from a height of 30 cm above the ground.

Both balls hit the ground together although their masses are different.

D–C

    a Why do they both hit the ground at the same time?

    ....

    .... **[1 mark]**

2 Sarah is a sky diver. She has a mass of 60 kg.

D–C

    a On the diagram mark and name the forces acting on Sarah as she falls.   **[2 marks]**

    b As Sarah leaves the aircraft, she starts to accelerate.

    Describe any difference in the size of the forces acting on her just after leaving the aircraft.

    .... **[1 mark]**

    c Sarah's acceleration decreases as she falls. Explain why.

    .... **[1 mark]**

    d Eventually she is travelling at terminal speed. Describe any difference in the size of the forces acting on her when she is falling at a constant speed.

    .... **[1 mark]**

    e Sarah opens her parachute. Explain what happens to each of the forces acting on her now?

B–A*

    ....

    .... **[2 marks]**

3 We often use the value 10 m/s^2 as the value for acceleration due to gravity on Earth. Describe how the acceleration due to gravity changes depending on where you are.

B–A*

    ....

    ....

    .... **[3 marks]**

The energy of games and theme rides

1 Finish the sentences. Choose words from this list.

gravitational potential energy (GPE)   **kinetic energy (KE)**

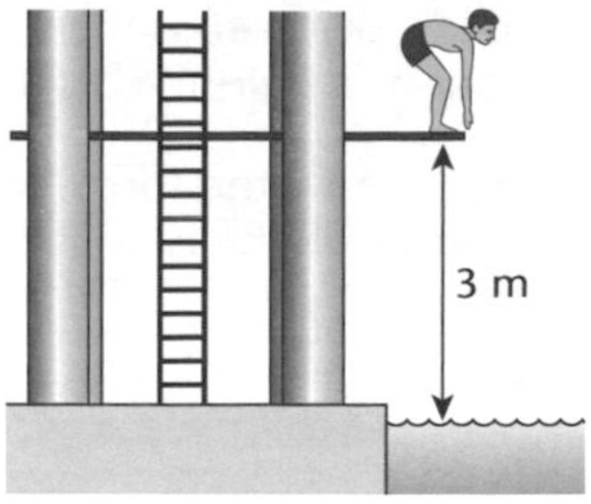

Rob is about to dive into the swimming pool.

He has .... As he dives ....

changes to .... Rob climbs to the 10 m board.

He has more .... than before.

[4 marks]

D–C

2 The diagram shows a roller coaster. The carriages are pulled up to B by a motor and then released.

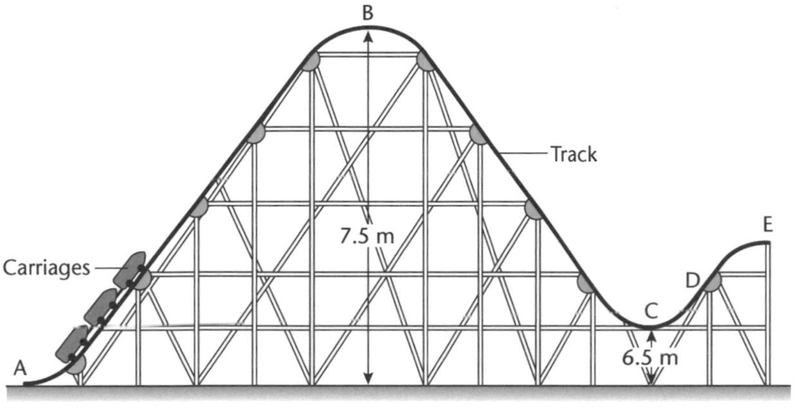

a At which point, A, B, C, D or E do the carriages have the greatest kinetic energy?

.... [1 mark]

D–C

b Describe the main energy change as the carriages move from B towards C.

.... [1 mark]

c Why must the height of the next peak at E be less than that at B?

....
.... [1 mark]

d The theme park decides to build a faster roller coaster. Suggest how they could modify the design to achieve this.

....
....
....
.... [3 marks]

3 49 000 years ago, an asteroid struck Earth and formed the Barringer crater in the Arizona desert. It is estimated the asteroid was travelling at 11 km/s when it struck. What is the equivalent height the asteroid fell from assuming it did not reach terminal speed and the effect of the Earth's atmosphere was negligible.

....
....
.... [2 marks]

B–A*

P3 Extended response question

Joe is a skydiver. He jumps from a plane at a height of 10 000m and free falls until his parachute opens at a height of 3000 m.

Explain how the forces acting on him affect the speed of his descent from the time he leaves the aircraft until he is about to land on the ground.

❗ The quality of written communication will be assessed in your answer to this question.

....
....
....
....
....
....
....
....
....
....
....
....
....
....
....
....
....
....
....
....
....
....
....
....
....
....
....
....
....
....
....
....
....

[6 marks]

Sparks

1 a Sally stands on an insulating mat. She puts her hands on the dome of an uncharged Van de Graaff generator. The Van de Graaff generator is switched on and Sally's hair starts to stand on end.

i Why does Sally stand on an insulating mat?

.... [1 mark]

ii Why must the Van de Graaff generator be uncharged when Sally puts her hands on it?

.... [1 mark]

iii What happens to Sally when the Van de Graaff generator is switched on?

.... [1 mark]

iv Why does this make Sally's hair stand on end?

.... [2 marks]

D–C

b Donna hangs two small, light plastic balls on nylon threads side by side.

She touches each ball with a charged polythene strip.

Why do the balls repel each other?

....

.... [2 marks]

c Paul does the same experiment but this time he only touches one ball with a charged polythene strip. The balls move towards each other but do not touch.

Give as much detail as you can to explain why this happens.

....

....

.... [2 marks]

B–A*

2 a Use your knowledge of electrostatics to explain the following.

i You sometimes get an electric shock on closing a car door after a journey.

....

.... [2 marks]

ii You should never shelter under a tree during a thunderstorm.

.... [1 mark]

iii Cling film often sticks to itself as it is unrolled.

.... [1 mark]

B–A*

b A factory uses machinery with moving parts. How does the machinery become charged?

....

.... [1 mark]

c Why do the operators at the factory stand on rubber mats?

....

.... [3 marks]

Uses of electrostatics

1 An electrostatic precipitator is placed in the chimney of a power station. It contains some wires and plates which are connected to a high voltage supply.

D–C

a If the wires are given a negative charge, what charge will be gained by the plates?

.... [1 mark]

b What charge do the soot particles gain as they pass close to the wires?

.... [1 mark]

B–A*

c Explain what has happened to the soot particles in terms of electrons.

.... [1 mark]

d What force is acting between the soot particles and the plates?

.... [1 mark]

2 In a bicycle factory the frames are painted using an electrostatic sprayer. The paint is positively charged. The frames are given the opposite charge to the paint.

nozzle is charged up positively

object to be painted is negatively charged

D–C

a Why does the paint spread out on leaving the sprayer?

....

.... [2 marks]

b Why are the bicycle frames given the opposite charge to the paint?

....

.... [2 marks]

c Lana decides to repaint her bicycle frame. She uses a paint spray that charges the paint positively but she does not charge the frame.

i What charge, if any, does Lana's bicycle frame acquire?

.... [1 mark]

B–A*

ii What problem does this cause?

.... [1 mark]

3 A defibrillator delivers an electric shock through the chest wall to the heart.

a What does the electric shock do to the heart?

.... [1 mark]

b The paddles of a defibrillator, charged from a high voltage supply, are placed on the patient's chest. Why should the chest be clean shaven and dry?

D–C

.... [2 marks]

c Why does the operator call out 'clear' before using the paddles?

.... [1 mark]

d A current of about 50 A passes through the patient for about 4 ms (0.004 s).

In general, such a large current would be fatal. Why can it be used in this situation?

.... [1 mark]

Safe electricals

1 a Adeela sets up the circuit shown. She then moves the variable resistor to include a greater length of wire into the circuit.

variable resistor

i What effect will this have on the brightness of the lamp?

.... [1 mark]

ii Add an ammeter and voltmeter to Adeela's circuit to allow her to measure the current through the lamp and the potential difference across the lamp.

[2 marks]

iii If the voltmeter reads 6 V and the ammeter 0.25 A, calculate the resistance of the lamp.

....

.... [3 marks]

D–C

b Adeela adjusts the variable resistor to 30 W and the current increases to 0.4 A.

i What is the voltage across the lamp now?

....

.... [3 marks]

ii Explain the change in resistance in terms of electron movement.

....

....

.... [3 marks]

B–A*

2 A battery has positive and negative terminals. Give two differences between the voltage from a battery and mains voltage.

....

.... [2 marks]

D–C

3 a i Label the live (L), neutral (N) and earth (E) wires in the plug shown. [3 marks]

ii Which wire, live, neutral or earth, is a safety wire?

.... [1 mark]

iii How does it work?

....

.... [2 marks]

b i An electric kettle passes a current of 10.5 A when working normally. Should the plug contain a 5 A or 13 A fuse?

.... [1 mark]

D–C

ii Why are fuses always connected in the live wire?

.... [2 marks]

iii The kettle is made from metal. How do the fuse and earth wire stop a person receiving an electric shock if they touch the kettle when it is faulty?

....

.... [4 marks]

Ultrasound

1 a Sound is a longitudinal wave.

i Explain how sound travels through the air to reach your ear.

....

.... [2 marks]

ii How does the frequency of a note change if its pitch increases?

.... [1 mark]

b What is 'ultrasound'?

.... [1 mark]

c Light is an example of a transverse wave. What is the difference between a longitudinal and a transverse wave?

....

.... [2 marks]

2 High-powered ultrasound is used to treat a patient with kidney stones.

a How does ultrasound do this?

....

.... [3 marks]

b Why must high-powered ultrasound be used?

.... [2 marks]

3 a Finish the sentences about an ultrasound scan.

A .... of ultrasound is sent into a patient's body. At each boundary

between different .... some ultrasound is ....

and the rest is transmitted. The returning .... are used to build up

an .... of the internal structure. A .... is placed

on the patient's body between the ultrasound .... and their

.... Without it nearly all the .... would be

.... by the .... [6 marks]

b Give two factors that affect the amount of ultrasound sent back to the detector at each interface within the body.

....

.... [2 marks]

c Air has a density of 1.3 kg/m³. Soft tissue has an average density of 1060 kg/m³.

i Explain why the gel used should have a density of about 1060 kg/m³.

.... [2 marks]

ii The time delay for an echo from ultrasound in soft tissue at a depth of 0.16 m was 0.2 ms (0.0002 s). Calculate the speed of ultrasound in soft tissue.

....

.... [4 marks]

D–C

D–C

B–A*

What is radioactivity?

1 a Complete the table showing the three types of nuclear radiation.

Type of radiation	Charge	What it is	Particle or wave
alpha			
beta			
gamma			

[3 marks] D–C

b Name the type of nuclear radiation that:

i does not change the composition of the nucleus .... [1 mark]

ii travels at about one-tenth the speed of light .... [1 mark]

2 a What is meant by 'half-life'? .... [1 mark]

b

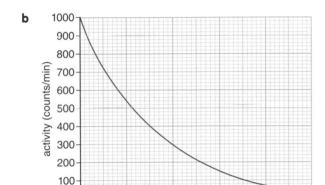

The graph shows how the activity of cobalt-60 changes with time. Use the graph to find the half-life of cobalt-60.

Show clearly, on the graph, how you got your answer. [2 marks] D–C

3 Radon, $^{220}_{86}$Rn, is radioactive. It decays to an isotope of polonium, $^{216}_{84}$Po with a half-life of 52 s.

a How many protons are there in a radon nucleus?

.... [1 mark]

b How many neutrons are there in a radon nucleus?

.... [1 mark]

c What is the name of the particle emitted in this decay?

.... [1 mark]

B–A*

d i Write a nuclear equation to describe the decay of a radon nucleus.

.... [3 marks]

ii Show that the atomic mass and mass number are conserved in this decay.

.... [2 marks]

e When a beta particle is emitted from a nucleus the mass number is unchanged but the atomic number increases by one. Explain how this is possible.

.... [2 marks]

f Finish the nuclear equation for the decay of iodine-131, emitting a beta particle.

$^{131}_{53}$I $\rightarrow$ $^{x}_{y}$Xe $+$ $^{0}_{-1}\beta$ x = ...   y = ... [2 marks]

Uses of radioisotopes

D–C

1 a Suggest two natural sources of background radiation.

.... [2 marks]

b Suggest two sources of background radiation that arise from human activity.

.... [2 marks]

D–C

2 Andrew works for an oil company. A leak has been reported in an underground pipe. He decides to locate the leak by introducing a small amount of radioisotope into the pipe.

a i What sort of radiation should the radioisotope emit?

.... [1 mark]

B–A*

ii Explain your choice. ....

.... [2 marks]

b What radiation detector could Andrew use? .... [1 mark]

D–C

c How will Andrew tell the site of the leak from his results?

.... [1 mark]

3 Smoke alarms use a source of alpha radiation in a small chamber.

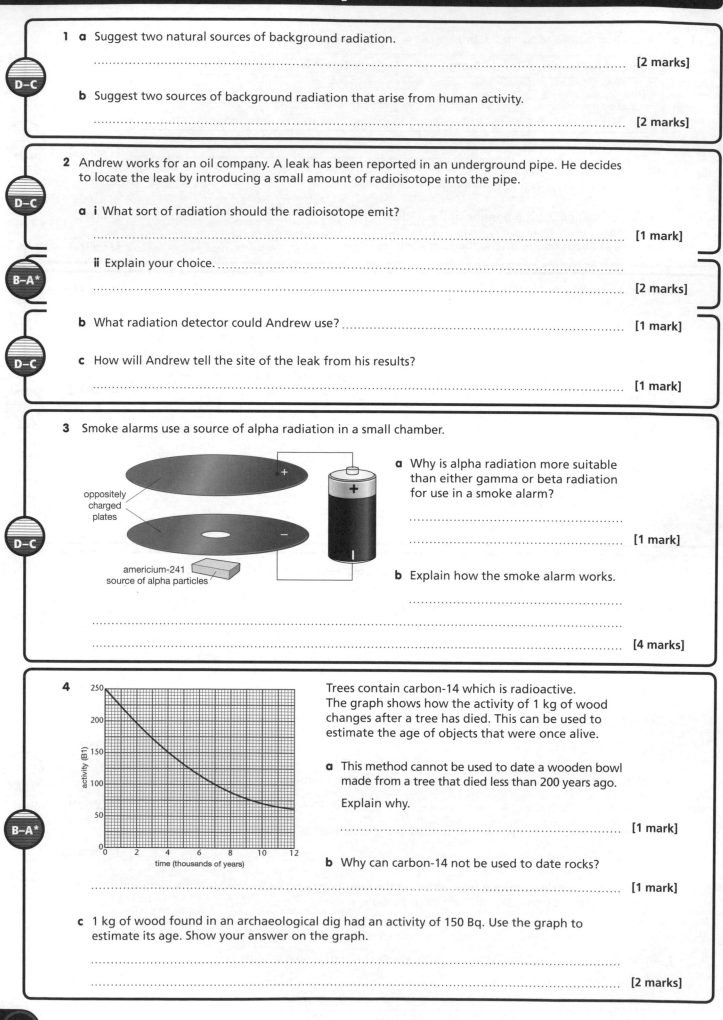

oppositely
charged
plates

americium-241
source of alpha particles

D–C

a Why is alpha radiation more suitable than either gamma or beta radiation for use in a smoke alarm?

....

.... [1 mark]

b Explain how the smoke alarm works.

....

....

.... [4 marks]

B–A*

4

activity (B1)

time (thousands of years)

Trees contain carbon-14 which is radioactive. The graph shows how the activity of 1 kg of wood changes after a tree has died. This can be used to estimate the age of objects that were once alive.

a This method cannot be used to date a wooden bowl made from a tree that died less than 200 years ago.

Explain why.

.... [1 mark]

b Why can carbon-14 not be used to date rocks?

.... [1 mark]

c 1 kg of wood found in an archaeological dig had an activity of 150 Bq. Use the graph to estimate its age. Show your answer on the graph.

....

.... [2 marks]

Treatment

1 a Why are x-rays and gamma rays suitable to treat cancer patients?

....

.... **[2 marks]**

b Why are alpha and beta particles not suitable to treat cancer patients?

....

.... **[1 mark]**

c Explain why nuclear radiation has to be used on patients very carefully.

....

.... **[3 marks]**

d How are radioisotopes for medical use produced?

.... **[1 mark]**

D–C

2 a What is a radioactive tracer? ....

.... **[2 marks]**

b Why is it used? ....

.... **[2 marks]**

c What sort of radiation should a tracer emit?

.... **[1 mark]**

D–C

d X-rays and gamma rays have similar properties. Why are gamma rays suitable to use as tracers but x-rays are not?

....

.... **[2 marks]**

B–A*

3 Three similar sources of radiation are used to destroy a brain tumour.

source of radiation

1

source of radiation

2

source of radiation

3

tumour

a i What type of radiation would be most suitable?

.... **[1 mark]**

ii Give a reason for your choice.

....

....

.... **[2 marks]**

b Why are there three sources of radiation arranged as shown?

....

....

.... **[2 marks]**

B–A*

c Describe an alternative technique to destroy a brain tumour that uses only one source of radiation.

....

.... **[2 marks]**

Fission and fusion

1 a Complete the following sentences explaining how a power station works.

The .... provides heat to boil the ....
to produce ....
The pressure of the .... turns the ....
which turns the .... making electricity. **[3 marks]**

b What is meant by 'fission'?

.... **[2 marks]**

c Enriched uranium is used as the fuel in a nuclear power station.
What is 'enriched uranium'?

.... **[2 marks]**

d A slow neutron can be captured by a uranium-235 nucleus and the 'new' nucleus then becomes unstable.

i Describe what happens after the capture of the neutron by the uranium nucleus.

....
.... **[3 marks]**

ii How can this one event become a continuous chain reaction?

.... **[1 mark]**

2 In one type of nuclear power station graphite is used as a moderator and boron rods are used to control the number of nuclear fissions in a given time.

a What does a 'moderator' do?

.... **[1 mark]**

b Why is a moderator necessary?

.... **[1 mark]**

c How do the boron rods control the number of fissions?

....
.... **[2 marks]**

3 a What is nuclear fusion?

.... **[1 mark]**

b Under what conditions does nuclear fusion take place in stars?

.... **[1 mark]**

c Why would fusion be preferable to fission as an energy source for the future?

....
.... **[2 marks]**

d Why are the claims that cold fusion is possible not widely accepted?

.... **[1 mark]**

P4 Extended response question

Doctors sometimes diagnose or treat a patient's illness without using surgery. They can use a number of different methods including x-rays, ultrasound or radioactive isotopes.

The doctor suspects that a female patient, who is pregnant, may have a tumour in her lung.

Suggest how the doctors should diagnose her condition and explain the reasons why this method is more suitable than the others.

! The quality of written communication will be assessed in your answer to this question.

[6 marks]

1 a How long does it take a geostationary satellite to orbit the Earth?

.... [1 mark]

b Polar orbiting satellites are used to help forecast the weather. Why are polar orbiting satellites used?

....

.... [1 mark]

c How does the height of a polar orbiting satellite compare with the height of a geostationary satellite?

.... [1 mark]

d How does the period of a polar orbiting satellite compare with the period of a geostationary satellite?

.... [1 mark]

e Why does a satellite in polar orbit travel faster than a satellite in geostationary orbit?

....

....

.... [2 marks]

2 a STENSAT is a communications satellite orbiting the Earth. Gravity is the centripetal force that keeps STENSAT in orbit. What is meant by a centripetal force?

....

.... [2 marks]

b The time taken for Mercury to orbit the Sun is less than the time taken for Venus to orbit the Sun. Explain why.

....

....

....

.... [2 marks]

c i Describe how the speed of a comet changes during one orbit of the Sun.

....

....

.... [2 marks]

ii Explain the reason for these changes.

....

....

....

.... [2 marks]

Vectors and equations of motion

1 a What is the difference between a scalar quantity and a vector quantity?

....

.... **[2 marks]**

b Force is one example of a vector.

i Give one other example of a vector.

.... **[1 mark]**

ii Give one example of a scalar.

.... **[1 mark]**

D–C

c David and Jean are both pushing their father's car in the same direction. David pushes with a force of 600 N and Jean with a force of 450 N. What is the resultant force on the car?

.... **[1 mark]**

d John is trying to open the door. He pushes with a force of 550 N. His friends, Jean and Erica are trying to stop him from opening the door. Jean pushes with a force of 300 N and Erica with a force of 250 N. What is the resultant force on the door?

.... **[1 mark]**

e Brian and Sophie are trying to move a heavy, square box. Brian pushes one side of the box and Sophie pushes on an adjacent side at right angles to Brian. Brian pushes with a force of 400 N and Sophie with a force of 300 N.

i What is the resultant force on the box?

.... **[1 mark]**

ii In which direction does the box move?

.... **[1 mark]**

B–A*

2 Tina is driving her car onto a motorway. She was travelling at 9 m/s as she accelerated along the slip road. Her acceleration was 3.6 m/s^2. She accelerated for 5 s until she reached the end of the slip road.

a Calculate her final velocity.

....

....

....

.... **[2 marks]**

D–C

b Calculate the length of the slip road.

....

....

....

.... **[2 marks]**

B–A*

Projectile motion

1 a Mark throws a ball horizontally as hard as he can from the top of a tall cliff. At the same time, Mary drops a similar ball from the same height. It takes 2.4 s for Mary's ball to hit the sea below.

i How long does it take for Mark's ball to hit the sea? Put a ring around the correct answer.

    less than 2.4 s   **exactly 2.4 s**   **more than 2.4 s**   **[1 mark]**

ii Mark throws the ball with a horizontal velocity of 4 m/s. What is the horizontal velocity when the ball hits the sea?

.... **[1 mark]**

b Nathan fires an arrow at a target. It falls just short. He cannot pull the bow string with any more force. What can he do to increase the range of his arrow?

....

.... **[1 mark]**

2 Newton suggested that if you could throw a ball with enough speed from the top of a tall tower, it would never come back to Earth but would stay in orbit. Explain why the ball would not come back to Earth.

....

....

.... **[2 marks]**

3 Freddie bowls a cricket ball horizontally from a height of 2 m at a velocity of 40 m/s (nearly 150 km/h). The batsman misses the ball and it lands about 5 m behind the wicket. What is the resultant velocity and direction of the ball when it hits the ground?

....

....

....

....

....

.... **[3 marks]**

Action and reaction

1 Steve is playing pool. His cue strikes the ball with a force of 100 N.

a What is the force of the ball on the cue?

D–C

.... [1 mark]

b The cue ball is travelling at 0.5 m/s when it hits another ball. The two balls move off together in the same straight line. What is their speed?

B–A*

....

....

.... [2 marks]

2 Aluminium-27 is an isotope of aluminium with a mass of 27 units. A stationary aluminium-27 atom is bombarded by a neutron, of mass 1 unit, travelling at 630×10^3 m/s. The neutron is absorbed and aluminium-28 is formed. Calculate the speed with which the aluminium-28 nucleus moves away.

B–A*

....

....

....

.... [2 marks]

3 a A mass of gas is trapped in a syringe. The plunger is free to move.

i What happens to the plunger when the syringe is warmed?

.... [1 mark]

ii Explain why this happens.

D–C

....

....

.... [2 marks]]

b Explain how the moving particles in the gas cause a pressure on the walls of the syringe.

....

....

B–A*

....

.... [3 marks]

4 The Saturn rocket used to send people to the Moon weighed 2.8 million kg at lift-off. Its engines delivered 34.5 million N of thrust. Explain how such a large force was created.

....

....

B–A*

....

.... [2 marks]

Satellite communication

1 a A geostationary satellite is used for communications. A television company transmits a microwave signal from its Earth station to a satellite. What happens to this signal once it is received by the satellite?

....

.... [2 marks]

D–C

b Explain how some radio waves are able to travel from one side of the Earth to the other without using a satellite.

....

....

....

.... [3 marks]

B–A*

c Why do satellite transmitting and receiving dishes need to be in careful alignment?

....

.... [2 marks]

2 Microwaves do not diffract as well as long-wave radio waves.

a Explain the difference between microwaves and long-wave radio signals.

....

.... [1 mark]

b With the aid of labelled diagrams, describe what is meant by diffraction.

B–A*

....

....

.... [2 marks]

c What are the optimum conditions for maximum diffraction?

....

.... [1 mark]

P5 Space for reflection

Nature of waves

1 Ray connects two loudspeakers to the same sound generator. He places them about 2 m apart on the laboratory bench. He stands in front of one loudspeaker and walks towards the other.

  a Describe what he hears as he walks from one loudspeaker to the other.

  ....

  ....

  .... **[2 marks]**

  D–C

  b Explain why he hears the sound as he does.

  ....

  ....

  ....

  ....

  .... **[4 marks]**

  B–A*

2 Kirsty arranges the apparatus shown below.

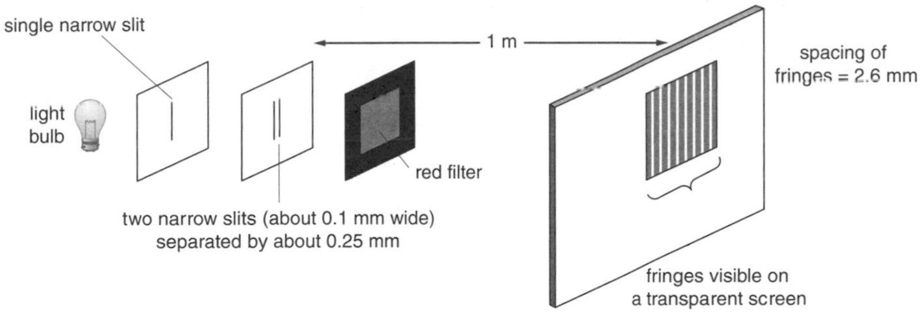

single narrow slit

light bulb

two narrow slits (about 0.1 mm wide) separated by about 0.25 mm

red filter

1 m

spacing of fringes = 2.6 mm

fringes visible on a transparent screen

  a Describe what Kirsty sees on the screen.

  D–C

  ....

  ....

  .... **[2 marks]**

  b Duncan wants to demonstrate polarisation of sound waves. His teacher tells him this is impossible. Why can Duncan not show polarisation of sound waves?

  ....

  ....

  .... **[2 marks]**

  c Polaroid is used to make sunglasses. Describe how it reduces the amount of light reaching the eye.

  B–A*

  ....

  ....

  .... **[2 marks]**

Refraction of waves

1 a Explain why light changes direction as it passes from air into glass.

.... [1 mark]

b The refractive index of amber is 1.55. Calculate the speed at which light travels through amber. (Speed of light in a vacuum = 300×10^6 m/s).

....

....

.... [2 marks]

2 Explain why a rainbow forms.

....

....

....

....

.... [3 marks]

3 a Describe one use of optical fibres.

....

.... [2 marks]

b Describe, with the aid of a diagram, how a fish, swimming below the surface of a pond, is able to see another fish swimming behind it by the process of reflection.

....

.... [2 marks]

c The refractive index of water is 1.33. What is the critical angle for light travelling from water into air?

....

....

.... [2 marks]

P5 Space for reflection

Optics

1 a Both the camera and the projector use lenses to produce an image. How are they adjusted to make sure the image is in focus?

.... [1 mark]

D–C

b Draw a ray diagram to show how the image is produced in:

i a camera

[4 marks]

B–A*

ii a projector.

[4 marks]

c Katie takes a photograph of a man 1.75 m tall. The image on the negative is 30 mm tall. Calculate the magnification.

....

....

.... [2 marks]

D–C

2 A convex lens is used as a magnifying glass. Explain how the image produced by a magnifying glass is different to the image produced by a projector.

....

....

....

.... [3 marks]

D–C

P5 Extended response question

In 1672, Isaac Newton published *Philosophical Transactions of the Royal Society* in which he attempted to prove, by experiment alone, that light consists of the motion of small particles he called corpuscles. At the same time, Robert Hooke and Christian Huygens were proposing a wave theory.

In 1803, Thomas Young addressed the Royal Society and produced compelling evidence that light is a wave.

Explain the evidence that has caused scientists to accept Young's theory and reject Newton's.

⚠ The quality of written communication will be assessed in your answer to this question.

....
....
....
....
....
....
....
....
....
....
....
....
....
....
....
....
....
....
....
....
....
....
....
....
....
....
....
....
....
....
....

[6 marks]

Resisting

1 Yasin builds the circuit shown below.

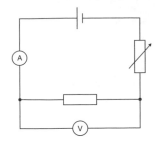

He wants to find out how the current and voltage change in a circuit when there is a resistor. He adjusts the variable resistor and records five sets of values for voltage and current. These are his results:

current in A	voltage in V
0.2	1.9
0.4	4.1
0.6	6.0
0.8	8.2
1.0	9.8

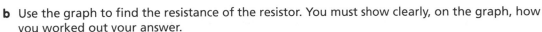

a i Plot the points on the grid. [2 marks]

ii Draw a line of best fit on the graph. [1 mark]

b Use the graph to find the resistance of the resistor. You must show clearly, on the graph, how you worked out your answer.

.... [2 marks]

2 a A bulb has the figures 230 V, 0.5 A printed on it. Calculate the resistance of the bulb.

.... [1 mark]

b When the bulb is turned on a dimmer switch is used to slowly increase the current through the bulb. How does the dimmer switch do this?

.... [1 mark]

c What happens to the atoms inside the filament when it becomes hot?

.... [1 mark]

d i Sketch a graph to show how the voltage across the bulb changes as the current increases.

[2 marks]

ii How can you find the resistance of the bulb from the graph?

.... [1 mark]

iii What happens to the resistance of a bulb when the current increases?

.... [1 mark]

Sharing

1 a The circuit shown is used to control the volume of a stereo. How can the volume be increased?

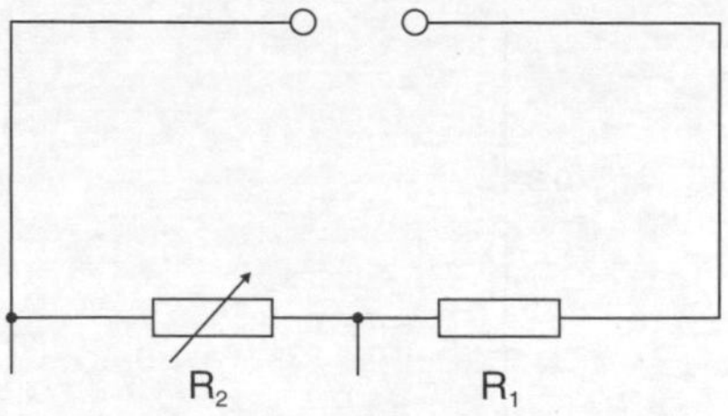

.... [1 mark]

b The supply voltage is 12V. $R_1 = 16\ \Omega$ and $R_2 = 8\ \Omega$. Calculate the voltage across R_2.

....

.... [2 marks]

2 a Nadaya is making a circuit in her electronics class. She must put 50 Ω in the circuit but she only has 100 Ω resistors. How can she connect the resistors to make 50 Ω?

....

.... [1 mark]

b Nadaya needs to reduce the resistance of her circuit to 25 Ω. How is this possible?

....

.... [1 mark]

3 a Some garden lights can come on automatically when it is dusk. What component is used in these lights to make this possible?

.... [1 mark]

b Draw the circuit which could be used to operate the garden lights. The gardener should be able to adjust the circuit so that it will activate at different threshold conditions.

[2 marks]

4 a How is the behaviour of a thermistor different to a resistor?

.... [1 mark]

b A room thermostat contains a thermistor in a potential divider circuit. Draw a circuit diagram that allows different temperatures to be set before the central heating switches on.

[2 marks]

It's logical

1

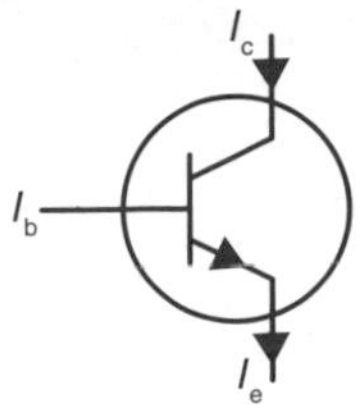

a Sandeep is using two transistors, like the one in the diagram, to make an AND gate. The base current is 5 mA and he is using this to switch on a current of 150 mA in the collector. What will the size of the emitter current be?

.... [2 marks]

b Why does Sandeep need two transistors to make his logic gate?

.... [1 mark]

c Complete the diagram to show how the AND gate is constructed.

[2 marks]

d Why is there a high value resistor in the base circuit?

.... [1 mark]

2 Work out truth tables for each of the following situations and state what type of logic gate could be used in the circuit.

a An alarm sounds when it has been armed and the front door is opened.

[1 mark]

b A courtesy light comes on when the passenger door or driver's door of a car is opened.

[1 mark]

c The warning light at the front of a washing machine lights if there is no water supply or water the heater is off.

[1 mark]

d The heater comes on in the conservatory if it is both dark and cold.

[1 mark]

3 a Draw a circuit diagram showing a potential divider circuit that can be used to switch on a logic gate when the temperature is hot.

[2 marks]

b Use ideas about resistance to explain how the logic gate is switched on when the temperature is hot.

....

.... [3 marks]

Even more logical

1 a Kashia connects the following logic circuit.

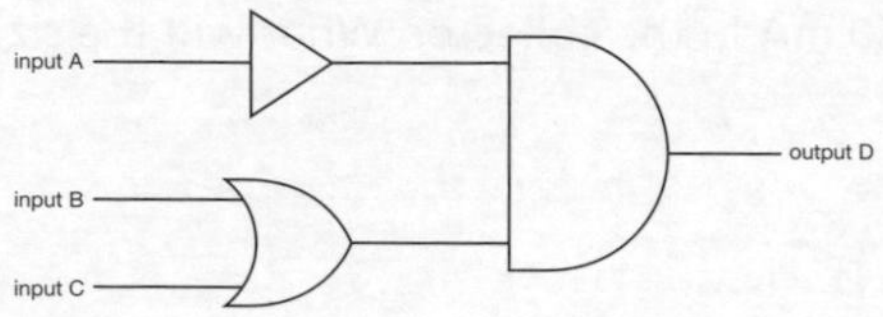

Construct the truth table for this circuit.

D–C

[4 marks]

b The top gate is replaced by an OR gate. Rewrite the truth table to show how the new circuit behaves.

B–A*

[4 marks]

2 Tammy connects a light emitting diode (LED) to the output from a logic gate. It lights.

a Suggest one use for a light emitting diode (LED) connected to the output from a logic gate.

.... [1 mark]

D–C

b Tammy uses a relay between the output from a logic gate and a heater for her greenhouse.

i Explain how the relay is used to switch on the heater.

....

....

.... [3 marks]

ii Why is the relay needed to switch on the heater?

B–A*

....

.... [2 marks]

Motoring

1 A wire is placed between the poles of a magnet. When a current is passed along the wire, the wire moves.

   a What happens to the wire if the current direction is reversed?

   .... **[1 mark]**

   b What happens to the wire if the poles of the magnet are reversed?

   .... **[1 mark]**

   c Draw a diagram to show the magnetic field around a straight wire. You must show the direction of the current and the direction of the field.

   [2 marks]

   d The diagram shows a wire passing between the poles of a magnet. Explain how the direction in which the wire moves can be predicted.

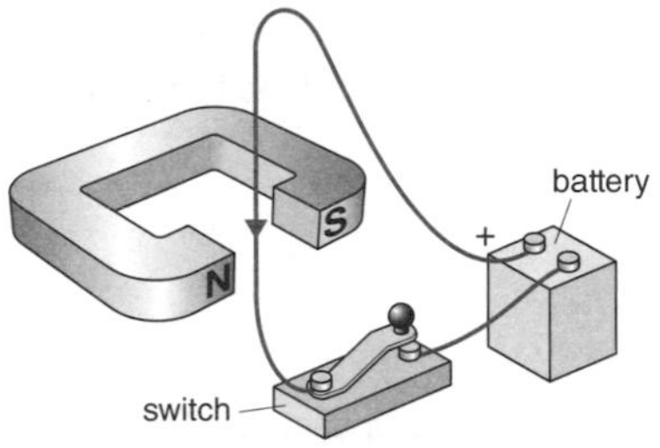

   ....
   .... **[3 marks]**

2 a A simple electric motor consists of a coil of wire rotating in a magnetic field. What three things can be done to make the motor spin faster?

   ....
   ....
   .... **[3 marks]**

   b Motors have commutators. What is the job of the commutator?

   ....
   .... **[1 mark]**

   c How is the magnetic field in a practical motor different to the model motor made in the classroom?

   ....
   .... **[1 mark]**

D–C

B–A*

D–C

B–A*

Generating

1 a Marcus connects the ends of a coil of wire to a sensitive ammeter. He holds the coil between the poles of an electromagnet. What does he see on the ammeter?

.... **[1 mark]**

b He switches the electromagnet off. What effect does this have on the ammeter?

.... **[1 mark]**

c He uses a variable resistor to increase the magnetic field in the electromagnet from zero to a maximum slowly. What effect does this have on the ammeter?

.... **[1 mark]**

d Marcus builds a model motor with the coil and connects it to an oscilloscope. Describe how the speed of spinning the coil affects the size and frequency of the voltage generated.

....

.... **[2 marks]**

2 a The bicycle dynamo contains a permanent magnet rotating inside a coil. The magnet is turned by the wheel of the bicycle rotating. What happens to the bicycle lamp when the cyclist stops at traffic lights?

.... **[1 mark]**

b The magnetic field in the generator at a power station is not produced by a permanent magnet. Describe how the field is produced.

....

.... **[1 mark]**

c How can the voltage produced by the generator be increased?

....

....

.... **[2 marks]**

d A turbine is used to turn the rotor coil. Why is it important that the rotor turns at an almost constant speed?

....

....

.... **[2 marks]**

e A DC generator has a commutator. What part of an AC generator delivers a current to the brushes?

.... **[1 mark]**

Transforming

1 Kate uses a transformer to make her door bell work from the mains supply. The bell works on 12 V. The number of turns on the primary coil of the transformer is 10 000. The number of turns on the secondary coil is 500.

a Explain how the construction of the transformer allows the mains voltage to be stepped down in this way.

....

.... [1 mark]

D–C

b Calculate the mains voltage that must be supplied to the primary coil.

....

....

....

.... [3 marks]

c Explain how an alternating current in the primary coil of a transformer produces an alternating current in the secondary coil.

....

....

.... [2 marks]

B–A*

d An isolating transformer has the same output voltage as the input voltage. Why use an isolating transformer as a shaving socket in the bathroom?

....

....

.... [2 marks]

2 a Electricity reaches our homes from the power stations through a network of cables called the National Grid. Why is it transmitted at high voltages in the cables?

....

.... [1 mark]

D–C

b Electricity is generated at 25 kV and transmitted at 400 kV.

i What difference will this make to the current in the overhead cables?

.... [2 marks]

B–A*

ii What effect will this have on the power?

.... [1 mark]

Charging

1 a Sketch a graph to show the current–voltage characteristics for a diode.

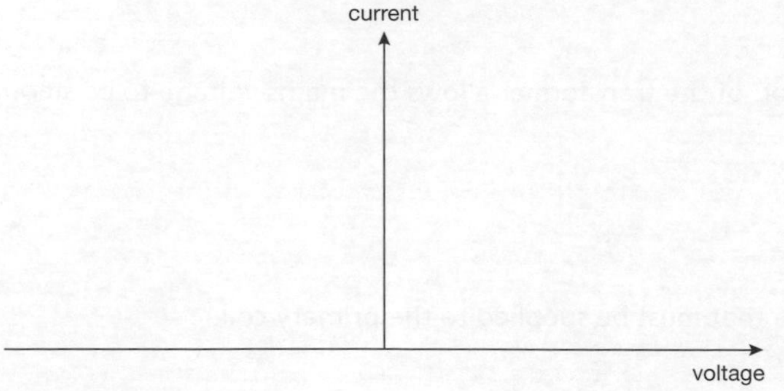

D–C

current

voltage

[1 mark]

b A diode consists of a piece of n-type semiconductor and a piece of p-type semiconductor joined together to form a junction. Use ideas about electrons to explain how a diode works.

....

....

....

.... [4 marks]

B–A*

c A single diode can produce half-wave rectification. For full-wave rectification a number of diodes are needed. Describe how diodes can be used to provide full-wave rectification. You should draw a diagram to help with your explanation.

[3 marks]

2 a Sketch a graph to show how the voltage across an uncharged capacitor changes as it is charged.

D–C

[2 marks]

b A capacitor is sometimes placed in a bridge rectifier circuit. What effect does this have on the output from the rectifier circuit?

.... [1 mark]

c Explain how this effect is achieved.

....

....

.... [3 marks]

B–A*

P6 Electricity for gadgets

When a mobile phone is plugged into the mains to be charged two things must happen.
First, the mains voltage must be reduced to a lower level suitable for the mobile phone battery.
Second, the voltage must be changed from AC to DC in order to charge the battery.

Describe how the each of these steps is achieved, including which components are needed and how these work to produce the required output.

❶ The quality of written communication will be assessed in your answer to this question.

....
....
....
....
....
....
....
....
....
....
....
....
....
....
....
....
....
....
....
....
....
....
....
....
....
....
....
....
....
....
....
....
....
....
.... [6 marks]

P1	
I can interpret data on rate of cooling and understand the consequences of the direction of energy flow.	✓
I can understand the concepts of specific heat capacity and specific latent heat and use the equations.	✓
I can explain how energy is transferred and how losses to the atmosphere can be reduced.	✓
I can interpret data on energy saving strategies and understand the importance of energy efficiency.	✓
I can describe the features of a transverse wave and determine wavelength and frequency from a diagram.	✓
I can understand that refraction is due to a wave travelling at different speeds in different materials.	✓
I can describe how light and the Morse code have been used for communication.	✓
I can describe how light behaves when its angle of incidence is below, above and equal to the critical angle.	✓
I can provide reasons for poor mobile phone reception.	✓
I can explain how scientists check each other's results by publishing their findings.	✓
I can describe how infrared signals carry information to control electrical and electronic devices.	✓
I can recall how the properties of digital signals led to the digital switchover for television and radio.	✓
I can describe how reflection and refraction of radio waves can be an advantage and can be a disadvantage.	✓
I can state the advantages and disadvantages of DAB broadcasts.	✓
I can recall the properties of P waves and S waves from an earthquake.	✓
I can interpret data about how sunscreen and skin tone can protect the skin from damage when sunbathing.	✓
I am working at grades D–C	✓

I can explain the difference between temperature and heat.	✓
I can explain why temperature does not change during a change of state.	✓
I can explain why the trapped air in cavity wall insulation further limits energy transfer through the walls.	✓
I can relate the design features of a house to the reduction in energy loss.	✓
I can describe the diffraction pattern for waves and say how this depends on the size of the gap.	∼
I can describe diffraction effects in telescopes and other optical instruments.	
I can explain advantages and disadvantages of using different electromagnetic waves for communication.	✓
I can explain how a laser beam is used in a CD player.	✓
I can explain how microwaves and infrared radiation transfer energy to materials.	✓
I can provide reasons for signal loss and describe how this loss can be reduced.	✓
I can explain how the signal from an infrared remote device uses digital signals to control electronic devices.	✓
I can describe the advantages of using digital signals and optical fibres for the rapid transmission of data.	✓
I can explain how long distance communication is achieved.	✓
I can explain why there is no interference when listening to digital radio.	✓
I can explain how seismic waves can be used to provide evidence for the structure of the Earth.	✓
I can describe how the ozone layer protects the Earth and the effects of its depletion.	✓
I am working at grades B–A*	

P2 Grade booster checklist

P2	
I can describe advantages and disadvantages of using photocells to produce electricity.	✓
I can describe advantages and disadvantages of wind turbines.	✓
I can describe how a simple alternating current generator works and how to increase the output.	✓
I can calculate the efficiency of a power station.	✓
I can explain how human activity and natural phenomena both have effects on weather patterns.	✓
I can distinguish between opinion and evidence based statements in the global warming debate.	✓
I can calculate the amount of electricity used in kilowatt hours and use this to find its cost.	✓
I can explain that transformers are used in the National Grid to reduce energy waste and costs.	✓
I can describe the relative penetrating power of alpha, beta and gamma radiations.	✓
I can describe some methods of disposing of radioactive waste.	✓
I can recall the relative sizes and nature of planets, stars, comets, meteors, galaxies and black holes.	✓
I can recall some difficulties of manned space travel and explain how information from space can be sent back to Earth.	✓
I can describe some of the evidence for past large asteroid collisions.	✓
I can describe how a collision between two planets could have resulted in the Earth-Moon system.	✓
I can recall that all galaxies are moving apart and that microwave radiation is received from all parts of the Universe.	✓
I can describe the end of the life cycle of small and large stars.	✓
I am working at grades D–C	✓

I can describe how light produces energy in a photocell and how to increase the current output.	✓
I can explain how passive solar heating works.	✓
I can recall that an efficient solar collector must track the position of the Sun in the sky.	✓
I can rearrange the formula to calculate efficiency in a power station.	✓
I can explain the greenhouse effect in detail.	✓
I can explain how scientists agree on the greenhouse effect but disagree on its causes.	✓
I can explain describe the advantages and disadvantages of using off peak electricity.	✓
I can explain how increasing the voltage in electricity transmission reduces energy waste through heat.	✓
I can describe experiments to show the relative penetrating powers of alpha, beta and gamma radiation.	✓
I can describe the advantages and disadvantages of nuclear power and explain the problems of dealing with radioactive waste.	✓
I can explain that gravitational attraction provides the centripetal force for orbital motion in our Solar system.	✓
I can explain why a light-year is a useful unit for measuring distances in space.	✓
I can explain why the speed of a comet changes as it approaches a star.	✓
I can suggest possible action that could be taken to reduce the threat of near-Earth objects.	✓
I can explain how the Big Bang theory accounts for red shift and how the age of the Universe can be estimated.	✓
I can explain the properties of a black hole.	✓
I am working at grades B–A*	✓

P3 Grade booster checklist

P3	
I can interpret the relationship between initial speed, final speed, distance and time.	✓
I can interpret the gradient of a speed–time graph.	✓
I can draw and interpret the shapes of speed–time graphs.	✓
I can describe the significance of positive and negative acceleration and calculate acceleration.	✓
I can describe and interpret the relationship between force, mass and acceleration.	✓
I can explain the factors that affect thinking distance, braking distance and their implications on road safety.	✓
I can calculate weight from knowledge of mass and gravitational field strength.	✓
I can calculate power knowing the time it takes to perform work.	✓
I can use and apply the equation KE $= \frac{1}{2}mv^2$.	✓
I can explain how electrically powered cars do cause pollution but not at their point of use.	✓
I can describe the relationships between momentum, mass, velocity, force and time.	✓
I can explain how and why crumple zones, seat belts and air bags reduce injuries.	✓
I can explain the motion of a falling object in terms of balanced and unbalanced forces.	✓
I can recognise that all objects fall with the same acceleration at a point on the Earth's surface.	✓
I can use the equation GPE = mgh and interpret examples of energy transfer from GPE to KE.	✓
I can interpret the energy transfers of a roller-coaster and the effects of mass and speed on kinetic energy.	✓
I am working at grades D–C	✓

P3	
I can interpret the relationship between speed, distance and time to include the effect of changing quantities.	✓
I can draw and interpret distance–time graphs.	✓
I can calculate acceleration and distance travelled from speed–time graphs.	✓
I can interpret the relationship between acceleration, speed change and time.	✓
I can use and manipulate the equation linking force, mass and acceleration.	✓
I can interpret and explain the shapes of graphs for thinking and braking distance against speed.	✓
I can perform a two-stage calculation linking work done and power.	✓
✳ I can explain the derivation of the equation power = force × speed.	
I can apply the ideas of kinetic energy to everyday situations such as braking distance.	✓
I can explain and evaluate the factors upon which fuel consumption depends.	✓
I can use Newton's second law of motion to explain the relationship between force, mass and acceleration.	✓
✳ I can evaluate car safety features and explain how ABS brakes work.	
I can explain why objects reach a terminal speed.	✓
I can understand that gravitational field strength varies with height and position on the Earth's surface.	✓
✳ I can understand the energy transfers when a body is falling at a terminal speed.	
I can use and apply the relationship mgh $= \frac{1}{2}mv^2$.	
✳ **I am working at grades B–A***	

P4 Grade booster checklist

P4	
I can state that like charges repel and opposite charges attract.	
I can describe electrostatic phenomena in terms of transfer of electrons which have a negative charge.	
I can describe problems and dangers caused by static electrical charge.	
I can explain how static electrical charge can be useful in dust precipitators, defibrillators, crop spraying and paint spraying.	
I can calculate resistance and describe how the resistance of a wire can be changed.	
I can describe the functions of the three wires and the fuse in a three pin plug.	
I can describe the features of a longitudinal wave.	
I can recall what ultrasound is and describe how it is used to scan the body and to break down kidney and other stones.	
I can explain and use the concept of half-life of radioactive isotopes.	
I can describe radioactivity as naturally occurring radiation from the nucleus of an unstable atom and give examples of sources of background radiation.	
I can recall the natures of alpha, beta and gamma radiation.	
I can describe how radioactive tracers are used in industry and in hospitals.	
I can describe other uses of radioactive isotopes such as in smoke detectors or radioactive dating.	
I can describe how x-ray images are produced.	
I can describe how nuclear fission is used in power stations and how the reaction is controlled.	
I can describe nuclear fusion and explain why it is difficult to produce power from it.	
I am working at grades D–C	

I can explain how problems with static electricity, including electric shocks, can be reduced.	
I can explain uses of electrostatics in terms of electron movement.	
I can rearrange the formula for resistance.	
I can explain the reasons for using fuses and circuit breakers in circuits.	
I can compare the motion of particles in longitudinal and transverse waves.	
I can explain how ultrasound is used in medicine.	
I can explain why ultrasound is used instead of x-rays for some scans.	
I can interpret graphical data on radioactive decay and half-life.	
I can construct and balance nuclear equations for alpha and beta decay.	
I can explain why alpha particles are so strongly ionising.	
I can evaluate the relative significance of sources of background radiation.	
I can explain how radiocarbon dating finds the age of old materials.	
I can explain how x-rays and gamma rays are produced.	
I can explain how radioactive sources are used to treat cancer.	
I can explain what causes a chain reaction.	
I can explain why 'cold fusion' is still not accepted by most scientists.	
I am working at grades B–A*	

P5 Grade booster checklist

P5	
I can explain how centripetal force keeps bodies moving in a circle.	
I can explain why satellites orbit at different heights and have different orbital periods.	
I can describe the difference between vectors and scalars and find the resultant of parallel vectors.	
I can use the equations $v = u + at$ and $s = \frac{1}{2}(u+v)t$.	
I can describe the horizontal and vertical motion of a projectile.	
I understand that the only force acting on a projectile is gravity and how this affects its vertical motion.	
I understand the principle of action and reaction between interacting bodies.	
I can use the kinetic theory to explain the relationship between volume, temperature and pressure.	
I can describe how electromagnetic waves behave differently as they transmit information.	
I can describe the wave patterns when plane waves pass through different sized gaps.	
I can describe interference in terms of reinforcement and cancellation.	
I can describe diffraction and polarisation of light.	
I can explain why refraction occurs, calculate refractive index and explain the reasons for dispersion.	
I can describe what happens when light is incident below, at and above the critical angle.	
I can describe the features of a convex lens and the effect the lens has on parallel and divergent beams of light.	
I can describe the action of a convex lens in a camera, projector and magnifying glass.	
I am working at grades D–C	

I can describe how the variation in gravitational force affects the orbits of planets and comets.	
I can explain why satellites in different orbits travel at different speeds and how they continually accelerate.	
I can calculate the resultant of two vectors that are at right angles to each other.	
I can use and apply the equations of motion.	
I understand that the resultant velocity of a projectile is the sum of its horizontal and vertical velocities.	
I can apply the equations of motion to projectiles.	
I can apply the conservation of momentum to colliding bodies.	
I can explain pressure in terms of change of momentum creating a force.	
I can explain why satellite transmitting and receiving aerials need very careful alignment.	
I can describe how the amount of diffraction is related to gap size and wavelength.	
I can explain why waves constructively and destructively interfere.	
I can explain the diffraction pattern for light and how polarisation is used to eliminate glare.	
I can interpret data on refractive indices and speed of light and relate this to dispersion.	
I can explain how the refractive index is related to the critical angle.	
I can explain the refraction of light as it passes through a convex lens and use ray diagrams to illustrate this.	
I can describe the properties of real and virtual images.	
I am working at grades B–A*	

P6 Grade booster checklist

P6	
I can explain the effect of a variable resistor in a circuit.	
I can describe how voltage-current graphs show the resistance of ohmic and non-ohmic devices.	
I can use the kinetic theory to explain resistance and temperature.	
I can explain how two resistors can be used as a potential divider and understand that the output depends on their relative values.	
I can describe how the resistance of an LDR and a thermistor can vary.	
I can understand that connecting resistors in parallel reduces the total resistance.	
I can describe benefits and drawbacks of miniaturisation of electronic components.	
I can understand that, in a transistor, the current in the emitter is found by adding the base current and the collector current.	
I can recall that logic gates are made from a combination of transistors and recognise the circuit diagram for an AND gate as two transistors.	
I can describe the truth tables for AND and OR gates or logic systems with three inputs.	
I can describe how a relay works.	
I can explain how a current carrying conductor in a magnetic field experiences a force and how this can be used in an electric motor.	
I can explain how the relative motion of a conductor and magnetic field produces electricity and how this is used to generate electricity.	
I can describe the construction of step-up, step-down and isolating transformers.	
I can recognise the current-voltage characteristic of a diode and explain how diodes can be used for rectification.	
I can describe the charging and discharging of a capacitor.	
I am working at grades D–C	

I can calculate the resistance of an ohmic conductor from a voltage–current graph.		
I can explain the shape of a voltage-current graph for a lamp, in terms of increasing resistance and temperature.		
I can explain how a potential divider can be constructed with an LDR or thermistor, or with variable resistors, to provide varying outputs.		
I can calculate the total resistance for resistors in parallel.		
I can complete a labelled diagram to show how a npn transistor can be used as a switch for an LED and explain why a high resistor is placed in the base circuit.		
I can describe the truth table for NAND and NOR logic gates and for systems with up to four inputs.		
I can explain how a thermistor or LDR can be used with fixed or variable resistors to provide an input for a logic gate.		
I can explain why a relay is needed for a logic gate to switch on a mains circuit.		
I can explain how Fleming's left-hand rule is used to predict the direction of the force on a current carrying wire in a magnetic field.		
I can explain how a commutator works and why practical motors have radial fields.		
I can explain how the size of an induced voltage depends on the rate at which the magnetic field changes.		
I can explain how an AC generator works including the action of the slip rings and brushes.		
I can describe how the changing field in the primary coil of a transformer induces an output in the secondary coil.		
I can explain how power loss in transmission of electricity is related to the current flowing in the transmission lines and explain why power is transmitted at high voltages.		
I can describe the action of a silicon diode in terms of the movement of holes and electrons and explain how four diodes can produce full-wave rectification.		
I can explain the action of a capacitor in a smoothing circuit.		
I am working at grades B–A*		

Index

P1 Energy for the home

Page 74 Heating houses

1 a Energy flows from a warm to a cooler body; temperature of the warmer body falls

2 Temperature is a measure of hotness on an arbitrary scale; heat is a form of energy on an absolute scale

3 A thermogram uses colours to represent different temperatures; the car engine/tyres/exhaust will be hot; a thermogram will show colours representing high temperature against the cold field OR

Thermal imaging cameras detect IR radiation; hot objects such as recently used car engines give out IR; the hotter the object the brighter (whiter) the image produced

4 a Specific heat capacity

b i Energy needed = mass × specific heat capacity × temperature change = 0.5 × 3900 × 70; = 136 500 J

ii Heat the beaker

5 a Specific latent heat

b Energy needed to break bonds; holding molecules together

Page 75 Keeping homes warm

1 a Particles in solid close together/particles in gas far apart/no particles in vacuum; gap between glass filled with gas or vacuum/more difficult to transfer energy than in solid

b i Air in foam is a good insulator/reduces energy transfer by conduction; air is trapped/unable to move; reduces energy transfer by convection

ii Energy from the room is reflected back in winter; energy from the Sun is reflected back in summer

2 a Particles are in constant motion and transfer kinetic energy – conduction; particles in solid close together so transfer energy easily/air is a gas so particles are far apart and it is more difficult to transfer energy

b Air expands when heated; density = mass + volume, so increased volume means less density

3 a Only 32% of energy input is useful as energy

b Output = 0.32 × 9.5; = £3.04

c Energy is lost up the chimney

Page 76 A spectrum of waves

1 a The maximum displacement of a particle from its rest position/allow clear labelling of diagram from axis to peak or trough

b The distance between two successive points having the same displacement and moving in the same direction/allow clear labelling of diagram showing this

c Number of complete waves passing a point each second

2 Wavelength = speed ÷ frequency;
$1500 \div 250\,000 = 6 \times 10^{-3}$ m; $1500 \div 125\,000 = 12 \times 10^{-3}$ m

3 Reflections from both mirrors; reflected ray returns along incident path

4 a Curved waves centres on gap; same wavelength

b Spikes or rings around the star

Page 77 Light and lasers

1 Dots and dashes series of on and off signals; not continuously variable signal

2 White light – many colours, different frequencies; laser light – one colour single frequency; laser light in phase

3 a i x angle of refraction greater than angle of incidence; y ray glancing along water/air boundary; z ray reflected back into water; angle of reflection = angle of reflection (by eye)

ii c marked as angle between ray and normal in diagram y

b Light down one set of optical fibres; reflected from internal organs; up a second set of fibres viewed by eyepiece or camera

Page 78 Cooking and communicating using waves

1 a Microwave radiation is more penetrating than infrared

b Microwaves need line of sight; there are no obstructions in space

2 a Gamma rays

b Wavelength of radiation from iron is longer than wavelength of radiation from element

(2 marks)

3 Radio waves diffracted around hill; short wavelength/ microwaves do not show much diffraction

Page 79 Data transmission

1 Greater choice of programmes; interact with programmes; information services

(Any 2)

2 3 – 5 reflections along length of fibre; ray reflected from surface with equal angles (by eye)

3 Interference on digital signals not evident; because signal is either high or low; many signals transmitted simultaneously/multiplexing; signals each divided up into short segments; transmitted signal takes segments from each, then recombined at end

P2 Answers

Page 80 Wireless signals

1 a Less refraction at higher frequencies

b Because it is aimed at a very small object

2 Reflected

3 a The foreign radio station is broadcasting on the same frequency; the radio waves travel further because of weather conditions

b There will be no interference from other stations

Page 81 Stable Earth

1 a Transverse – S and longitudinal – P; travels through solid – P and S; travels through liquid – P

b Waves refracted by core; cause shadow on opposite side of Earth

c S waves do not pass through liquid; not detected on opposite side of Earth

2 a 20 × 15; 300 minutes /5 hours

b Repeat their readings; consult other scientists about their findings

3 Newspaper may not be reliable source; based on only one piece of evidence; no peer review

Page 82 P1 Extended response question

5–6 marks

A detailed description of how microwave radiation requires line of sight and how it is affected by hills (little diffraction) and water (scattering the signals) A discussion of possible dangers from use of mobile phone near the head and that children are more at risk because their bodies are still developing.

All information in answer is relevant, clear, organised and presented in a structured and coherent format. Specialist terms are used appropriately. Few, if any, errors in grammar, punctuation and spelling.

3–4 marks

A limited description of some of the details of how microwave signals are transmitted and affected by hills and water. A mention of possible dangers from microwaves and children more at risk.

For the most part the information is relevant and presented in a structured and coherent format. Specialist terms are used for the most part appropriately. There are occasional errors in grammar, punctuation and spelling.

1–2 marks

An incomplete description, stating that microwave radiation is used and signal strength is affected by hills and water. A mention of possible dangers from microwaves.

Answer may be simplistic. There may be limited use of specialist terms. Errors of grammar, punctuation and spelling prevent communication of the science.

0 marks

Insufficient or irrelevant science. Answer not worthy of credit.

P2 Living for the future

Page 83 Collecting energy from the Sun

1 a Photocells do not need fuel; do not need cables; need little maintenance; use a renewable energy source; the operation of photocells cause no pollution or global warming

(Any 4)

b n-type silicon has an impurity added to produce an excess of electrons and p-type silicon has a different impurity to produce an absence of free electrons

(3 marks)

2 a Increase the light intensity; increase the surface area exposed to light; decrease the distance to the light source

(Any 2)

3 a During the night the walls and floor radiate infrared at a longer wavelength and this is reflected by the glass so it stays inside the room

b i S at left end of infrared

ii P at right end of infrared

4 a The speed of the wind

b Advantages – renewable; no pollution or global warming

Disadvantages – unreliable; noisy; spoil the landscape

(Any 2)

Page 84 Generating electricity

1 a Use a stronger magnet; increase the number of turns in the coil; spin the coil faster

(Any 2)

b In a power station the turbine spins the generator, electromagnets provide field, electromagnets turn inside coil where current is induced

c i Indicates one complete cycle on time axis

ii 300 V

2 a Turbine turns generator

b 60 MJ

Page 85 Global warming

1 a By absorbing the longer wavelength infrared radiation

b Water vapour

2 a Increased it

b Natural forest fires; volcanic eruptions; decay of dead plant and animal matter; escape from the oceans; respiration

(Any 4)

c The mining and burning of fossil fuels; cattle farming; rice paddies; the burying of waste in landfills

(Any 3)

P2 Answers

3 a Sun's radiation has a short wavelength which is absorbed by the Earth. The Earth then re-radiates but with a longer wavelength. Longer wavelength radiation is absorbed by the greenhouse gases

(3 marks)

b Increase – The smoke from factories reflects radiation from the towns back to Earth

Decrease – Ash clouds from volcanoes reflect radiation from the Sun back into space

(1 mark each)

4 a They disagree on how much humans are contributing to global warming

b On the basis of scientific evidence

c Polar ice caps melting; rising sea levels

5 a Fact

b Opinion

c Fact

Page 86 Fuels for power

1 a 24 W *(2 marks)*

b 15p *(3 marks)*

c Less demand but supply still available

d 30 minutes

2 a Availability; ease of extraction; effect on the environment; associated risks

(Any 3)

b The advantages and disadvantages of each type of power station; the availability and ease of obtaining each type of fuel; the energy output of each type; the costs of building and maintaining each type; any sensible suggestion

(Any 3)

3 a Reduces energy loss; reduced cost

b The current goes down as the voltage goes up

c This leads to less heating in the wires and reduced energy losses

Page 87 Nuclear radiations

1 a False; true; true; false

b i Negative – radiation causes atom to gain one or more electrons

(2 marks)

Positive – radiation causes atom to lose one or more electrons

(2 marks)

ii Ionisation can lead to changes/mutations in DNA/ protein molecules may change shape; this mutation can lead to cancer

2 a Radioactive tracer; radiotherapy

(Any 1)

b Penetrates the body easily; can kill mutated cells

(Any 1)

3 The radiation ionises the oxygen and nitrogen atoms in air; this causes a very small electric current that is detected; when smoke fills the detector in the alarm during a fire the air is not so ionised; the current is less and the alarm sounds

4 a Waste can remain radioactive for thousands of years

b Radioactive waste is not suitable for making nuclear bombs; it could be used by terrorists to contaminate water supplies; or areas of land; or to frighten the public

(Any 3)

Page 88 Exploring our Solar System

1 a i A star is a ball of hot glowing gas giving out energy from fusion

ii A planet is a spherical object orbiting a star, which has cleared the neighbourhood of its orbit

iii A meteor is made from grains of dust that burn up as they pass through the Earth's atmosphere

b i A galaxy is a collection of stars

ii A black hole is are formed when a large star dies, you cannot see a black hole because light cannot escape from it

c i The centripetal force caused by gravity

ii Arrow on Moon towards centre of circle

2 a To prevent astronaut being the blinded by the Sun's glare

b Planet – Mars

Explanation – Mars is the only planet near enough to the Earth for humans to travel to

3 The distance light travels in one year

Page 89 Threats to Earth

1 a Mars; Jupiter

b The gravitational field of Jupiter is too strong and prevents this

(2 marks)

c Unusual metals have been found near craters; fossils are found below the metal layer but not above; fossil layers have been disturbed by tsunamis

(Any 2)

2 The average density of Earth is 5500 kg/m³ while that of the Moon is only 3300 kg/m³; there is no iron in the Moon; the Moon has exactly the same oxygen composition as the Earth

(Any 2)

3 a C on elliptical orbit

b X nearest the Sun

c Solar wind pushes it away from Sun

4 a So that we could track their position accurately; and deal with any that came too near Earth

b i Send a rocket with explosives near the NEO; and detonate the explosives to alter its path

P3 Answers

ii Be careful not to split asteroid into pieces which may still hit Earth; be careful that explosion is far enough away not to disrupt Earth

Page 90 The Big Bang

1 a The ones furthest away

b Light spectra from distant galaxies are shifted to the red end of the spectrum

(3 marks)

c The greater the red shift the faster the galaxy is moving

2 a Cloud pulled together by gravity; star becomes smaller, hotter and brighter; the core temperature is hot enough for nuclear fusion to take place; fusion releases energy

b First becomes a red giant; gas shells, called planetary nebula, are thrown out; the core becomes a white dwarf; then cools to become a black dwarf

3 a He observed the planets' orbits with a telescope

b Because the Church disagreed with the ideas

c Newton's law of gravitation/ Newton used gravity to explain the orbit of the planets

Page 91 P2 Extended response question

5–6 marks

A detailed description of how the Sun started as a cloud of gas and dust pulled together by gravitational forces. When the temperature becomes high enough thermonuclear fusion starts to join hydrogen nuclei together to form helium nuclei. This main sequence phase continues for about 10 billion years until the hydrogen runs out and the star cools and expands to form a red giant. Next the outer layers called planetary nebulae are ejected and the core shrinks to form a white dwarf and then cools to a black dwarf. Includes the idea that our Sun was formed from the remains of a supernova remnant.

All information in answer is relevant, clear, organised and presented in a structured and coherent format. Specialist terms are used appropriately. Few, if any, errors in grammar, punctuation and spelling.

3–4 marks

A limited description of some of the details of our Sun's life cycle. Includes the idea that our Sun was formed from the remains of a supernova remnant.

For the most part the information is relevant and presented in a structured and coherent format. Specialist terms are used for the most part appropriately. There are occasional errors in grammar, punctuation and spelling.

1–2 marks

An incomplete description, may simply list the stages. Mentions the fact that the Sun was formed from the death of another star.

Answer may be simplistic. There may be limited use of specialist terms. Errors of grammar, punctuation and spelling prevent communication of the science.

0 marks

Insufficient or irrelevant science. Answer not worthy of credit.

P3 Forces for transport

Page 92 Speed

1 a 80 s

b 80 m

c D to E

d Speed = $\frac{40}{40}$; = 1 m/s

e Speed = $\frac{40}{80}$; = 0.5 m/s

2 a Average speed = 12.5 m/s; time = $\frac{5000}{12.5}$; = 400 s

b Average speed = $\frac{500}{35}$; = 14.3 m/s

c Distance = 4500 m time = 365 s;

average speed = $\frac{4500}{365}$; = 12.3 m/s

Page 93 Changing speed

1 a

Uniform increase in speed; constant speed of 10 m/s; uniform decrease in speed; constant speed of 5 m/s

b Area under graph

2 a

b Change in speed = 15 – 3 = 12 m/s;

deceleration = $\frac{12}{15}$; = 0.8 m/s^2

c Distance travelled = $\frac{1}{2}$ (15 × 15) + (25 × 15) + $\frac{1}{2}$ (20 × 15); = 637.5m

3 Change in speed = acceleration × time; = 6 × 5 = 30 m/s is final speed

P3 Answers

Page 94 Forces and motion

1 a Force = $500 \times \frac{40}{20}$; = 1000 N

b Acceleration = $\frac{1250}{500}$; = 2.5 m/s^2

2 a Tiredness; age; under influence of drugs/alcohol; distracted/lacking concentration

(Any 2)

b i Thinking distance will increase

ii Reaction time is unchanged; so she will travel a greater distance in that time

c Less tread/less grip; braking distance increases

3 Thinking distance – straight line

Braking distance – curve showing 'squared' relationship; passing through (30, 13.5), (40, 24), (50, 37.5)

Page 95 Work and power

1 a Work done = 80×2; = 160 J

b i Height = $80 \times \frac{2}{60}$; = 2.7 m

ii Power = $80 \times \frac{2}{1.5}$; = 106.7 W

2 a Chris is more powerful than Abi

b Weight = 10×60; 600 N

c Power = $600 \times \frac{3}{8}$; = 225 W

d Power = $600 \times \frac{3}{12}$; 150 W

e 26 N/kg

3 a A

b C

c Fuel pollutes the environment; car exhaust fumes are harmful; carbon dioxide is a greenhouse gas; carbon dioxide contributes to climate change

(Any 3)

Page 96 Energy on the move

1 a Fewer road junctions; fewer speed changes; fewer gear changes

(Any 2)

b Fuel used = $\frac{96}{24}$; 4

c Renault Megane has smaller engine capacity

d Excessive acceleration and deceleration; speed changes; braking; driving in too low a gear

(Any 3)

2 a Recharging requires electricity from power stations which do cause pollution

b Advantage – no pollution; no batteries; does not need energy from power station

(Any 1)

Disadvantage – not always sunny; not a constant energy source

(Any 1)

3 a Kinetic energy = $\frac{1}{2} \times 1200 \times (20)^2$; = 240 000 J

b KE proportional to v^2; so braking distance is quartered not halved

Page 97 Crumple zones

1 a Seat belt – stretches so that kinetic energy is transferred into elastic potential; crumple zones – absorb some of car's KE by changing shape on impact; air bag – absorbs some of person's KE by squashing up around them

b Computer controls pressure on brakes; brakes are pumped; prevents locking and skidding; increases braking force just before skid

(Any 2)

2 a Force = $\frac{(25 \times 55)}{0.5}$; = 2750 N

b Force = $\frac{(25 \times 55)}{0.002}$; = 687 500 N

3 a Force = mass × acceleration; to reduce force acceleration (deceleration) must be reduced since mass cannot change

b Crumple zone concertinas; to reduce the car's speed more slowly

Page 98 Falling safely

1 Acceleration is independent of mass

2 a Weight acting vertically downwards; air resistance acting vertically upwards

b Weight greater than air resistance

c The faster she falls the greater the air resistance

d Balanced/equal in size but opposite in direction

e Weight is unchanged; air resistance increases suddenly

3 Greater at poles than equator; increases down a mine; decreases up a mountain; different in space or another planet/moon

(Any 3)

Page 99 The energy of games and theme rides

1 a GPE; GPE; KE; GPE

2 a C

b GPE to KE

c Energy is transferred as sound/heat/friction

d Increase height of B to increase GPE; GPE is transferred to KE as the carriage falls; more KE means faster speed

3 $h = \frac{(11\,000)^2}{(2 \times 10)}$; = 6 050 000 m OR 6050 km

Page 100 P3 Extended response question

5–6 marks

A detailed description of how the forces affect the speed of descent to include; weight being greater than air resistance at the start so speed increases; air resistance increases with speed; terminal (constant) speed when air resistance equals weight; air resistance greater than weight when parachute opens so speed decreases; the parachute provides a much larger surface area and displaces more air molecules; air resistance decreases when slowing down; lower terminal speed when air resistance is again equal to his weight until he lands.

All information in answer is relevant, clear, organised and presented in a structured and coherent format. Specialist terms are used appropriately. Few, if any, errors in grammar, punctuation and spelling.

3–4 marks

A limited description of how the forces affect the speed of descent at two or three stages.

For the most part the information is relevant and presented in a structured and coherent format. Specialist terms are used for the most part appropriately. There are occasional errors in grammar, punctuation and spelling.

1–2 marks

An incomplete description of how the speed changes as he falls but with little if any reference to the forces acting.

Answer may be simplistic. There may be limited use of specialist terms. Errors of grammar, punctuation and spelling prevent communication of the science.

0 marks

Insufficient or irrelevant science. Answer not worthy of credit.

P4 Radiation for life

Page 101 Sparks

1 a i So that charge will build up on her body instead of flowing away to earth

   ii Otherwise she would get a shock

   iii Charges will begin to build up on her body

   iv Sally's hair all picks up the same charge; so it repels and stands up

 b Both balls will pick up electrons from the polythene and become negatively charged; since they have the same charge they will repel

 c The touched ball becomes charged; this induces the opposite charge on (one side) of the other ball; then the balls will attract due to these oppositely charged sides/one being charged and one uncharged; they do not touch as the force of attraction is weaker than if both balls were charged

(Any 2)

2 a i You become charged up by the friction of the rubber tyres on the road; when you touch the metal body it discharges

   ii A tree could be the highest point and the charge is more likely to hit that point and jump from the tree into you

   iii When you unroll cling film it becomes charged by the friction and may then attract to the uncharged parts

 b The moving parts can become charged by friction if they are insulators

c The charge could travel to earth through the operators; causing a spark; rubber mats will prevent this

Page 102 Uses of electrostatics

1 a Positive

 b Negative

 c They have gained electrons

 d Electrostatic attraction

2 a Paint particles all have same charge; so repel each other

 b To attract the paint; so there is less wasted paint

 c i Positive

   ii The frame will start to repel the paint

3 a The shock restores a regular heart rhythm

 b Hair is a poor conductor; water may conduct the charge across the chest away from where it is needed

 c To avoid any other people receiving a shock

 d It is safer as it is only for an extremely short time

Page 103 Safe electricals

1 a i The brightness reduces

   ii Ammeter in series; voltmeter added in parallel across the bulb

   iii $R = \dfrac{V}{I}$ OR $\dfrac{6}{0.25}$; $= 24$; Ω

 b i 30×0.4; 12; V

   ii There is a longer length of wire in the circuit now; the electrons have to pass through more atoms; so there are less electrons passing per second

2 Battery is DC/mains is AC; mains is high voltage/battery is low voltage

3 a i N on left; E at top; L on right

   ii Earth

   iii If case becomes live; current will flow down the earth wire; protecting the user

(Any 2)

 b i 13 A

   ii The live wire is the one at high voltage; so the fuse is placed in the live wire to break the circuit as close to this as possible

   iii If a fault develops where the live wire touches the case; the case becomes 'live'; a large current flows in the earth and live wires; and the fuse blows/melts breaking the circuit

Page 104 Ultrasound

1 a i A series of compressions/high pressure areas; rarefactions/low pressure areas in the air

   ii Frequency increases

P4 Answers

b Sound above 20 000 Hz

c In a longitudinal wave the vibrations of the particles are parallel to the direction of the wave; in a transverse wave the vibrations of the particles are at right angles to the direction of the wave.

2 a Ultrasound vibrations; pass into the body to the stones; the vibrations break up the stones

b High-powered ultrasound carries more energy; stones need more energy to break them up

3 a Pulse; tissues; reflected; echoes; image; gel; probe; skin; ultrasound; reflected; skin
(2 correct = 1 mark; all correct = 6 marks)

b Density of the tissues; speed of ultrasound in the tissues

c i The gel must have a similar density to the tissue; so that the ultrasound will pass through; very little will then be reflected at the boundary
(Any 2)

ii $\frac{0.0002}{2} = 0.0001$ sec; speed $= \frac{0.16}{0.000}$; $= 1600$; m/s

Page 105 What is radioactivity?

1 a

type of radiation	charge	what it is	particle or wave
alpha	+2	2 protons + 2 neutrons	particle
beta	−1	electron	particle
gamma	0	short wavelength electromagnetic radiation	wave

(1 mark each correct line or column; 3 marks total)

b i Gamma

ii Beta

2 a The time taken for the activity of a sample to drop to half of the original value; time for half of the original nuclei to decay
(Any 1)

b $5\frac{1}{2}$ years; lines drawn on graph to show this

3 a 86

b 134

c Alpha

d i $^{220}_{86}\text{Rn} \rightarrow {}^{216}_{84}\text{Po} + {}^{4}_{2}\text{He}/{}^{4}_{2}\alpha$
(All correct = 3 marks)

ii Atomic number 86 = 84 + 2; mass number 220 = 216 + 4

e A neutron changes into a proton increasing the atomic number; both are still in the nucleus so the mass number is unaffected

f $x = 131$; $y = 54$

Page 106 Uses of radioisotopes

1 a Rocks/soil; cosmic rays

b Radioactive waste from power stations; nuclear weapons testing; radioisotopes for medical uses
(Any 2)

2 a i Gamma

ii Gamma can penetrate the pipe and soil under the surface; alpha and beta would not penetrate the pipe or soil layer

b A Geiger counter; Geiger Muller GM tube; ratemeter/scalar counter
(Any 1)

c The count rate would drop after the site of the leak

3 a Alpha radiation will be stopped by the smoke particles/beta and gamma would not

b Without smoke, the alpha particles ionise the air; creating a tiny current that can be detected by the circuit in the smoke alarm

With smoke, the alpha particles are partially blocked so there is less ionisation of the air; the resulting change in current is detected and the alarm sounds

4 a The activity would not have dropped enough for the difference to be measured

b Carbon-14 is used for dating objects that were once living

c 4000 years; lines shown on graph

Page 107 Treatment

1 a They are both ionising radiations; they can kill cancer cells

b They cannot penetrate into the body to the site of the cancer

c Due to their ionising abilities; they can ionise the atoms in the DNA of normal cells; and cause damage to the patient's normal cells

d They are placed into a nuclear reactor/made to absorb neutrons

2 a A radioactive isotope which is introduced to the body; to diagnose a problem

b Tracers will travel around the body to the site of a problem; they can be detected outside the body; this avoids having to cut the patient open
(Any 2)

c Gamma

d Gamma rays are emitted by radioisotopes which can be introduced into the body; X-rays (are produced by high speed electrons hitting a metal target which cannot be done inside the body)

3 a i Gamma

ii It can penetrate to the site of the tumour; it can kill the cancerous cells

b This way the normal surrounding cells only receive one third of the radiation; the tumour receives the whole dose

c The source can be rotated around the patient; delivering a constant dose of radiation to the tumour and intermittent dose to the surrounding tissue

P5 Answers

Page 108 Fission and fusion

1 a Source of energy; water; steam; steam; turbines; generator

(1 mark for 2 correct words)

b Fission is the breaking down of a large nucleus; into smaller ones

c Uranium; that has had extra neutrons added

d i The uranium nucleus splits; releasing energy; and more neutrons

ii The extra neutrons released are captured by more uranium nuclei causing them to split and release even more neutrons

2 a Slows down the neutrons

b So that the neutrons are more likely to be captured by the uranium nuclei

b The control rods can be raised or lowered to absorb more or less neutrons; controlling how many neutrons are available for capture by the uranium nuclei

3 a The joining together of two lighter nuclei to make one heavier one

b Very high temperatures and pressures

c There is a plentiful supply of hydrogen/fuel (in sea water); waste products of fusion are less harmful than fission

d The experimental results cannot be reproduced by other scientists

Page 109 P4 Extended response question

5–6 marks

A detailed description of how ultrasound could produce an image including its reflection from different tissue boundaries, how the reflection depends on tissues having different densities and the different speed of the ultrasound in the different media, ultrasound echoes being used to build up the picture of her lungs and detect the tumour without the need for surgery.

The alternative methods, i.e. X-rays and gamma radiation cause ionisation in cells and are potentially harmful especially to the foetus whose cells are rapidly dividing. Ultrasound is safe as a scanning method even with unborn babies.

All the information in the answer is relevant, clear, organised and presented in a structured and coherent format. Specialist terms are used appropriately. There are few, if any, errors in grammar, punctuation and spelling

3–4 marks

A limited description of how the ultrasound method produces the image, lacking in specific details.

Insufficient detail in the comparison of ultrasound with alternative methods.

For the most part the information is relevant and presented in a structured and coherent format. Specialist terms are used for the most part appropriately. There are occasional errors in grammar, punctuation and spelling.

1–2 marks

An incomplete description of how ultrasound produces an image, with little if any reference to the advantages it may have.

Answer may be simplistic. There may be limited use of specialist terms. Errors of grammar, punctuation and spelling prevent communication of the science.

0 marks

Insufficient or irrelevant science. Answer not worthy of credit.

P5 Space for reflection

Page 110 Satellites, gravity and circular motion

1 a 24 hours

b View whole of Earth's surface

c Polar orbiting satellites are lower than geostationary

d Period of polar orbiting satellite is less than geostationary

e Gravitational force greater; larger centripetal acceleration

2 a Force acting towards centre of circle; keeping body moving in a circle

b Mercury has shorter distance to travel; moves at a greater speed (or reverse argument)

c i Speeds up on approach to Sun; slows down as it moves away

ii Gravitational force varies

(1 mark)

OR gravitational force greater when nearer Sun

(2 marks)

Page 111 Vectors and equations of motion

1 a Scalar has size only; vector has size and direction

b i Velocity; acceleration; momentum; displacement

(Any 1)

ii Length; volume; speed; mass; density; time; temperature; energy; work; power; pressure

(Any 1)

c 1050 N

d 0 N

e i 500 N

ii 37° to 400 N force

2 a $v = 9 + (3.6 \times 5)$; 27 m/s

b $s = 0.5 \times 36 \times 5$; 90 m

Page 112 Projectile motion

1 a i Exactly 2.4 s

ii 4 m/s

b Fire arrow at an angle above the horizontal but not above 45°

2 a Ball falls to Earth; but Earth curves away beneath ball

3 $v^2 = 2 \times 10 \times 2$ or $v = \sqrt{40}$;
resultant velocity = $\sqrt{(40 + 40^2)}$ or 40.5 m/s;
angle = $\tan^{-1}\sqrt{40}/40$ or 9° to horizontal

Page 113 Action and reaction

1 a 100 N

b 0.5 m = 2 mv; v = 0.25 m/s

2 $630 \times 10^3 \times 1 = 28$ v; v = 22.5×10^3 m/s

3 a i Plunger moves out

ii Particles move faster; hit the wall more frequently causing greater force

b Particle colliding with wall has momentum, mv and rebounds with momentum, –mv;
momentum change 2 mv; force = $\frac{\text{momentum change}}{\text{time}}$
leads to pressure = $\frac{\text{force}}{\text{area}}$

4 Large number of gas particles; moving at very high speed

Page 114 Satellite communication

1 a Amplified; retransmitted back to Earth

b Reflected from ionosphere; by total internal reflection; reflected off water/land surface

c Little diffraction; narrow beam

2 a Microwaves have a shorter wavelength

b Spreading out of a wave as it passes through a gap or around an obstacle;

c Gap size is similar size to wavelength of wave

gap is small, waves diffract

Page 115 Nature of waves

1 a Sound becomes louder and quieter; and louder again

b When sound is louder – two peaks meet/waves reinforce/constructive interference; whole wavelength path difference; when sound is quieter – peak meets trough/waves cancel out/destructive interference; odd half wavelength path difference

2 a Alternate bands/stripes/bars/fringes of red light; darkness

b Sound is not a transverse wave; only transverse waves can be polarised

c Light waves vibrate in all directions at right angles to wave direction; Polaroid only allows waves to vibrate in one direction

Page 116 Refraction of waves

1 a Speed changes

b Speed = $\frac{300 \times 10^6}{1.55}$; speed = 193×10^6 m/s

2 Sunlight is made up of different colours each of which travels at a different speed; each colour has a different refractive index

3 a Endoscope; to see inside body without surgery/ communications; transmission of laser light signals

b Angle of incidence greater than critical angle;

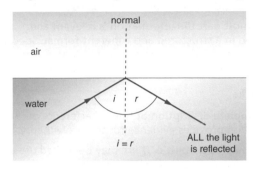

c Critical angle = $\sin^{-1}(\frac{1}{1.33})$; = 48.8°

Page 117 Optics

1 a Distance from lens to screen/film adjusted

b i Object/ one ray path through camera/ second ray path through camera/ image

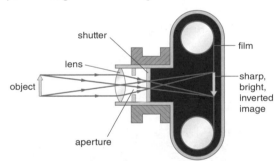

ii Light source and mirror/one ray path through projector/second ray path through projector/ image

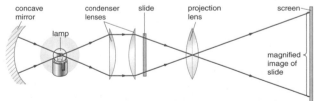

c Magnification = $\frac{30}{1750}$; = 0.017

2 Magnifying glass image is virtual, not real; cannot be projected onto screen; upright not inverted

(Any 3)

Page 118 P5 Extended response

5–6 marks

A detailed description of both interference and polarization to include the mechanisms for constructive and destructive interference and plane polarisation.

A description of how the path difference between two coherent light sources determines whether interference is constructive or destructive. The effect on light of shining through crossed polaroid filters is mentioned.

All information in answer is relevant, clear, organised and presented in a structured and coherent format. Specialist terms are used appropriately. Few, if any, errors in grammar, punctuation and spelling.

3–4 marks

A limited description of how interference and polarization provide evidence for the wave theory in that particles could not interfere or be polarised.

For the most part the information is relevant and presented in a structured and coherent format. Specialist terms are used for the most part appropriately. There are occasional errors in grammar, punctuation and spelling.

1–2 marks

An incomplete description of how interference or polarization provides evidence for the wave theory.

Answer may be simplistic. There may be limited use of specialist terms. Errors of grammar, punctuation and spelling prevent communication of the science.

0 marks

Insufficient or irrelevant science. Answer not worthy of credit.

P6 Electricity for gadgets

Page 119 Resisting

1 a i Points plotted correctly

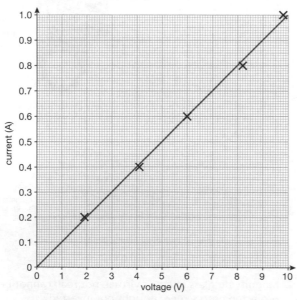

(3–4 correct = 1 mark; 5–6 correct = 2 marks)

ii Correct straight line through origin drawn with a ruler

b Gradient triangle shown on graph; 10 Ω

2 a $(R = \frac{V}{I}) \frac{230}{0.5} = 460$ Ω

b When the dimmer switch is turned a greater length of wire is added to the circuit

c The atoms vibrate more when the filament is hot

d i Increasing gradient; passing through origin

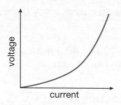

ii From instantaneous values of voltage and current

iii Increases

Page 120 Sharing

1 a Increase the resistance of the variable resistor (R_2)

b $(V_{out} = \frac{R_2}{R_1 + R_2})$ $V_{out} = 12 \times \frac{8}{(16 + 8)}$; 4V

2 a Connect two 100 Ω resistors in parallel

b Connect four 100 Ω resistors in parallel

3 a Light dependent resistor

b

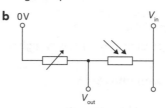

4 a Resistance of a thermistor decreases with temperature

b Potential divider circuit; variable resistor and thermistor correctly connected

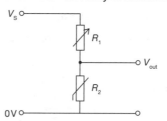

Page 121 It's logical

1 a $I_e = I_c + \frac{1_b}{1_e} = 150 + 5$; $= 155$ mA

b AND gate has two inputs

c

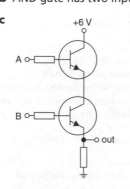

(2 marks)

d The base current should be low

P6 Answers

2 a AND gate

Alarm set	Door open	Alarm sounds
0	0	0
0	1	0
1	0	0
1	1	1

b OR gate

Driver's door	Passenger door	Courtesy light
0	0	0
0	1	1
1	0	1
1	1	1

c NAND gate

Water in	Heater on	Warning light
0	0	1
0	1	1
1	0	1
1	1	0

d NOR gate

Light	heat	Heater on
0	0	1
0	1	0
1	0	0
1	1	0

3 a Potential divider circuit; thermistor in correct place

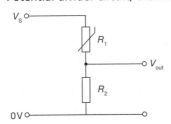

b Resistance of thermistor decreases when hot; output of potential divider becomes high; output is connected directly to input of logic gate

Page 122 Even more logical

1 a

A	B	C	D
0	0	0	0
1	0	0	0
0	1	0	1
0	0	1	1
1	1	0	0
1	0	1	0
0	1	1	1
1	1	1	0

b

A_1	A_2	B	C	D
0	0	0	0	0
1	0	0	0	0
0	1	0	0	0
0	0	1	0	0
0	0	0	1	0
1	1	0	0	0
1	0	1	0	1
1	0	0	1	1
0	1	1	0	1
0	1	0	1	1
0	0	1	1	0
1	1	1	0	1
1	1	0	1	1
1	0	1	1	1
0	1	1	1	1
1	1	1	1	1

2 a Indicator to show that something is on (or WTTE)

b i Output current from gate passes through an electromagnet; attracts an iron armature; this pivots and pushes an insulating bar against the contacts

ii Output current from logic gate is too low to power heater; and the higher current could damage the logic gate so the relay isolates it

Page 123 Motoring

1 a Moves in opposite direction

b Moves in opposite direction

c Concentric circles; direction correct relative to current

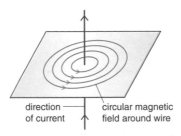

d Fleming's left hand rule. First finger- field; second finger- current; thumb – motion
(max 2 marks if left hand not mentioned)

2 a Increase current; increase number of turns on coil; increase magnetic field strength

b Reverses the current in the coil every half turn

c Practical motor has radial field

P6 Answers

Page 124 Generating

1 a Reading remains at zero

   b Current is produced

   c Less current is produced

   d Size of current greater if speed greater (ORA); frequency of AC greater if speed greater (ORA)

2 a Light goes off

   b By electromagnets

   c Increase the number of turns on the electromagnets; increasing the magnetic field strength

   d Maintain constant voltage; constant frequency

   e Via slip rings attached to the ends of the coil

Page 125 Transforming

1 a More coils on primary/fewer coils on secondary

   b $\frac{N_p}{N_s} = \frac{V_p}{V_s}$; $V_p = 12 \times \frac{10\,000}{500}$; $= 240$ V

   c Alternating current in the primary coil produces a changing magnetic field in the core; this induces a changing voltage in secondary coil

   d Output terminals are not live; there is no danger of electrocution if you touch the socket with wet hands

2 a To reduce energy loss from heat in cables

   b i Current is reduced; to $\frac{1}{16}$th of original value

   ii The power loss will be $\frac{1}{16^2}$ or $\frac{1}{256}$th of what it would have been

Page 126 Charging

1 a

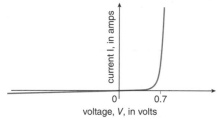

   b Formation of depletion layer; depletion layer acts as insulator; voltage applied with positive to n-type depletion layer widens no current; voltage applied with positive to p-type depletion layer narrows; increasing voltage depletion layer disappears and current passes

(Any 4)

   c Bridge circuit; input and output on opposite sides; diode directions correct

2 a Increasing gradient; from the origin

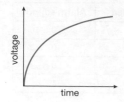

   b Smooths output to make it more constant

   c When the DC voltage from the rectifier circuit falls, the capacitor supplies current; capacitor charges quickly near peak of varying DC

Page 127 P6 Extended response

5–6 marks

A detailed description of how a step-down transformer is used to reduce the voltage from the mains, to include the fact that the primary (or input) coil has more turns than the secondary (or output) coil in order to reduce the voltage.

A description of a bridge rectification circuit, including four diodes (or circuit diagram) with details of how this would change the AC to DC. A smoothing circuit containing a capacitor to smooth out the output for charging the battery.

All information in answer is relevant, clear, organised and presented in a structured and coherent format. Specialist terms are used appropriately. Few, if any, errors in grammar, punctuation and spelling.

3–4 marks

A limited description of the transformer as well as the rectification circuit and the capacitor.

For the most part the information is relevant and presented in a structured and coherent format. Specialist terms are used for the most part appropriately. There are occasional errors in grammar, punctuation and spelling.

1–2 marks

An incomplete description of how the output voltage can be reduced and changed from AC to DC. Answer may be simplistic. There may be limited use of specialist terms. Errors of grammar, punctuation and spelling prevent communication of the science.

0 marks

Insufficient or irrelevant science. Answer not worthy of credit.